Application of Evolutionary Algorithms for Multi-objective Optimization in VLSI and Embedded Systems

M.C. Bhuvaneswari
Editor

Application of Evolutionary Algorithms for Multi-objective Optimization in VLSI and Embedded Systems

Springer

Editor
M.C. Bhuvaneswari
Electrical and Electronics Engineering
PSG College of Technology
Coimbatore, Tamil Nadu, India

ISBN 978-81-322-1957-6 ISBN 978-81-322-1958-3 (eBook)
DOI 10.1007/978-81-322-1958-3
Springer New Delhi Heidelberg New York Dordrecht London

Library of Congress Control Number: 2014947350

Printed on acid-free paper

Springer is part of Springer Science+Business Media (www.springer.com)

Preface

This book describes how evolutionary algorithms (EA) such as genetic algorithms (GA) and particle swarm optimization (PSO) can be used for solving multi-objective optimization problems in the area of embedded and VLSI systems design. This book is written primarily for practicing CAD engineers and academic researchers who wish to apply evolutionary techniques and analyze their performance in solving multi-objective optimization problems in VLSI and embedded systems.

Many real engineering problems have multiple objectives, such as minimizing cost, maximizing performance, and maximizing reliability. Being a population-based approach, EA are suitable for solving multi-objective optimization problems. Research has been carried out to ascertain the capabilities of GA and PSO in solving complex and large constrained combinational optimization problems. A reasonable solution to a multi-objective problem is to consider a set of solutions, each of which satisfies the objectives at an acceptable level without being dominated by any other solution. Graph theoretic approaches and integer/linear programming have been used to solve problems in embedded and VLSI system design. GA and PSO may prove to be a general purpose heuristic method for solving a wider class of engineering and scientific problems.

Organization of the Book

This book provides an introduction to multi-objective optimization using meta-heuristic algorithms: GA and PSO and their application to problems like hardware/software partitioning in embedded systems, circuit partitioning in VLSI, design of operational amplifiers in analog VLSI, design space exploration in high level synthesis, delay fault testing in VLSI testing, and scheduling in heterogeneous distributed systems. It is shown how in each case the various aspects of the EA, namely its representation, and operators like crossover, mutation, etc., can be separately formulated to solve these problems.

The content of this book is divided into 9 chapters, and each chapter focuses on one aspect or application of EA to multi-objective optimization. Each of the chapters deals with experimental results as well as an analysis of the problem at hand.

Chapter 1 provides an introduction to multi-objective GA and PSO algorithms. The terminology of GA and PSO is introduced and its operators are discussed. Also, the variants of GA and PSO and their hybrids with hill climbing are explained.

Chapter 2 addresses the problem of hardware/software partitioning in embedded systems. It discusses how multi-objective EA can be applied to hardware/software partitioning.

Chapter 3 focuses on the problem of circuit partitioning in VLSI. It describes how the chromosome is represented in GA for the problem of circuit partitioning.

Chapter 4 explains how the design of operational amplifiers in analog VLSI can be represented as multi-objective optimization problem and how GA can be applied to solve the design of operational amplifiers.

Chapter 5 describes how multi-objective EA can be applied to the problem of design space exploration in high-level synthesis of VLSI.

Chapter 6 presents the design space exploration of datapath (architecture) in high level synthesis (HLS) for computation intensive applications.

Chapter 7 elaborates evolutionary algorithm driven high level synthesis design flow: Algorithm to RTL.

Chapter 8 deals with delay fault testing in VLSI testing. It describes how the problem of delay fault testing can be formulated as a multi-objective problem and delineates experimental results for the various crossover operators.

Chapter 9 explores how the scheduling in heterogeneous distributed system can be formulated as a multi-objective problem and the applicability of EA to the problem. Experimental results are provided for four variants of multi-objective GA and PSO.

Coimbatore, India M.C. Bhuvaneswari

Acknowledgements

I would like to acknowledge the significant contributions of Ms. M. Shanthi, Dr. M. Jagadeeswari, Dr. S. Jayanthi, Dr. D. S. Harish Ram, and Dr. G. Subashini, who have contributed much to the research on application of multi-objective optimization in VLSI and embedded systems as part of their doctoral research. I would like to thank Dr. Anirban Sengupta for providing the chapters on Design Space Exploration of Datapath (Architecture) in High Level Synthesis for Computation Intensive Applications and Design Flow from Algorithm to RTL using Evolutionary Exploration Approach. I am grateful to the reviewers Dr. P. Navaneethan, Dr. Reza Javaheri, and Ms. Jayashree Saxena whose insightful comments enabled us to improve the quality of the book significantly. I am thankful to the editorial team at Springer, especially Swati Meherishi and Kamiya Khatter, for their support in publishing this book.

Finally, I am thankful to my family and friends for their support and encouragement.

M.C. Bhuvaneswari

Contents

About the Editor

M.C. Bhuvaneswari is Associate Professor in Department of Electrical and Electronics Engineering at PSG College of Technology, Coimbatore, India. She completed her Ph.D. in VLSI Design and Testing from Bharathiar University in 2002. She has over 20 years of research and teaching experience. Her areas of research interest include VLSI design and testing, genetic algorithms, digital signal processing, digital systems design and microprocessors. She has published her work in several journals and conferences of national and international repute. Dr. Bhuvaneswari is a Life Member of Indian Society for Technical Education, Institution of Engineers (India), Computer Society of India and System Society of India. She was honoured with the Dakshinamoorthy Award for Teaching Excellence in the year 2010.

Chapter 1
Introduction to Multi-objective Evolutionary Algorithms

M.C. Bhuvaneswari and G. Subashini

Abstract Many real engineering optimization problems can be modeled as multi-objective optimization problem (MOP). These problems actually do have multiple objectives that conflict each other, and optimizing a particular solution with respect to a single objective can result in unacceptable results with respect to the other objectives. Evolutionary algorithms (EAs) have emerged as flexible and robust meta-heuristic methods for solving optimization problems, achieving the high level of problem-solving efficacy spanning every application domain. The main motivation for using EAs to solve MOPs is because EAs deal simultaneously with a set of possible solutions. This allows us to find several members of the Pareto optimal set in a single run of the algorithm. Research works have ascertained that genetic algorithm (GA) and particle swarm optimization (PSO) find more suitability for solving complex and large constrained combinational optimization problems in engineering and scientific domains. Evolutionary techniques for multi-objective optimization are categorized into three approaches: plain aggregating approaches, population-based non-Pareto approaches, and Pareto-based approaches. This chapter provides an introduction to few, multi-objective evolutionary algorithms (MOEAs) based on GA and PSO. The algorithms described are based upon plain aggregating approaches and Pareto-based approaches. An insight into the several variants of the basic algorithm and how these algorithms could be hybridized with techniques like Hill Climbing for better performance is also described.

Keywords Multi-objective optimization • Evolutionary algorithms • Nondominated sorting • Genetic algorithm • Particle swarm optimization • Pareto hill climbing

---

M.C. Bhuvaneswari
Electrical and Electronics Engineering, PSG College of Technology, Coimbatore, India
e-mail: mcb@eee.psgtech.ac.in

G. Subashini (✉)
Information Technology, PSG College of Technology, Coimbatore, India
e-mail: suba@ity.psgtech.ac.in

M.C. Bhuvaneswari (ed.), *Application of Evolutionary Algorithms for Multi-objective Optimization in VLSI and Embedded Systems*, DOI 10.1007/978-81-322-1958-3_1,

1.1 Introduction

Many complex engineering optimization problems can be modeled as multi-objective formulations. In many real-life problems, objectives under consideration conflict each other and optimizing a particular solution with respect to a single objective can result in unacceptable results to other objectives. Many real engineering problems do have multiple objectives, that is, they minimize cost, maximize performance, maximize reliability, and so on. Being a population-based approach, evolutionary algorithms (EAs) are well-suited to solve multi-objective optimization problems (MOPs). Research works have been carried out to ascertain the capabilities of genetic algorithm (GA) and particle swarm optimization (PSO) in solving complex and large constrained combinational optimization problems. A reasonable solution to a multi-objective problem is a set of solutions. Each solution in the set satisfies the objectives at an acceptable level without being dominated by any other solution. Evolutionary multi-objective optimization methods are not yet widely used for problem solving in very large scale system integration (VLSI) and embedded systems. Graph theoretic approaches, integer/linear programming have been used to solve problems in embedded and very large scale integration system design. GA and PSO may prove to be a general purpose heuristic method for solving a wider class of engineering and scientific problems.

This chapter provides an introduction to multi-objective GA and PSO algorithms. The terminology of GA and PSO is introduced and its operators are discussed. Also the variants of multi-objective GA and PSO and its hybrids with Hill Climbing are discussed.

1.2 Multi-objective Optimization Problem

A multi-objective optimization problem (MOP) usually has no single optimal solution but a set of optimal trade-off solutions as opposed to a single-objective optimization problem (Deb 2001). An important task in multi-objective optimization is to identify a set of optimal trade-off solutions (called a Pareto set) between the conflicting objectives, which helps to gain a better understanding of the problem structure and supports the decision-maker in choosing the best compromise solution for the considered problem. An exhaustive study of multi-objective optimization methods for engineering problems is present in literature (Marler and Arora 2004).

In mathematical terms, the multi-objective optimization problem can be defined as in Eq. 1.1:

$$\text{Minimize/Maximize}\,\mathrm{F}(x) = [f_1(x), f_2(x) \ldots f_k(x)] \tag{1.1}$$

where $f_1(x)$, $\mathrm{f}_2(x) \ldots f_k(x)$ are the objective functions. The objective function can be a minimization or maximization type. A perfect multi-objective solution

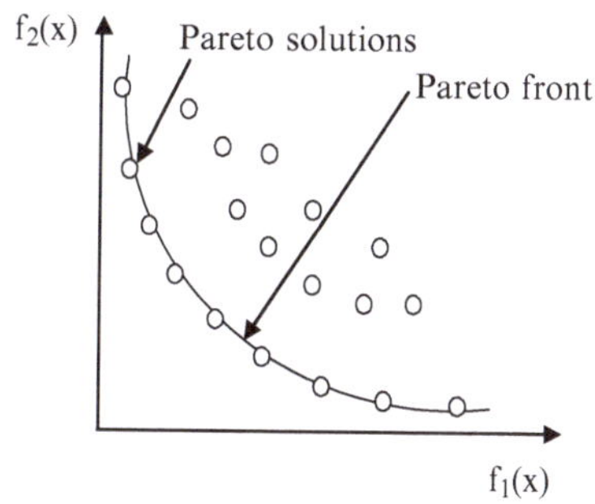

Fig. 1.1 Pareto front for two objective functions

that simultaneously optimizes each objective function is almost impossible. A reasonable solution to a multi-objective problem is to investigate a set of solutions, each of which satisfies the objectives at an acceptable level without being dominated by any other solution.

If all objectives are to be minimized, a feasible solution x is said to dominate another feasible solution y if and only it is as good as the other solution and better in at least one objective as given by Eqs. 1.2 and 1.3:

$$f_i(x) \leq f_i(y), \forall i \in [1, 2, \ldots, K] \text{ and} \tag{1.2}$$

$$f_j(x) < f_j(y), \exists j \in [1, 2, \ldots, K] \tag{1.3}$$

A solution is said to be Pareto optimal if it is not dominated by any other solution in the solution space. The set of all feasible nondominated solutions is referred to as the Pareto optimal set and the objective function values corresponding to the solutions in the Pareto set constitute the Pareto front (Deb 2001). A set of solutions for two objective functions where $f_1(x)$ and $f_2(x)$ are minimized is shown in Fig. 1.1.

1.3 Why Evolutionary Algorithms?

Evolutionary algorithms (EAs) are search methods that take their inspiration from natural selection and survival of the fittest in the biological world. EAs differ from more traditional optimization techniques because they involve a search from a population of solutions and not from a single point. Evolutionary algorithms are particularly suitable to solve multi-objective optimization problems as they consider a set of possible solutions (population). This finds several members of the Pareto optimal set in a single run of the algorithm instead of performing a series of separate runs as in the case of the traditional mathematical programming techniques. Also, evolutionary algorithms are less susceptible to the shape or continuity of the Pareto front. They can easily deal with discontinuous or concave Pareto front, whereas these two issues are a real concern for classical mathematical programming techniques (Kaisa 1999).

1.4 Multi-objective Evolutionary Algorithms

The ultimate goal of a multi-objective optimization algorithm is to identify solutions in the Pareto optimal set. Conventional optimization techniques such as gradient-based methods and simplex methods are designed considering a single objective. This requires the multi-objective optimization problem to be reformulated as a single-objective problem. This formulation results in a single best solution on each run when multiple solutions are possible.

Evolutionary algorithms seem particularly suitable to solve multi-objective problems, because they deal simultaneously with a set of possible solutions (Coello 1999; Kaisa 1999). Few traditional MOP techniques focus on search and others on multicriteria decision-making. However, Multi-objective evolutionary algorithms (MOEAs) are attractive solution techniques to MOP as they deal with both search and multicriteria decision-making (Zhou et al. 2011; Guliashki et al. 2009). Recently, there has been an increasing interest in applying EA to MOP. Also, Multi-objective evolutionary algorithms have been successfully applied in various researches to find good solutions for complex MOPs. Fonseca and Fleming (1995) categorize the evolutionary techniques for multi-objective optimization into three approaches: plain aggregating approaches, population-based non-Pareto approaches, and Pareto-based approaches.

In plain aggregating approach, all the objectives are often artificially combined, or aggregated, into a scalar function ultimately making the problem single-objective before optimization and EA is applied. Though this method has the advantage of producing a single compromise solution, new runs of the optimizer may be required until a suitable solution is found if the final solution cannot be accepted. This is caused due to an inappropriate setting of the coefficients of the combining function or the function excluding few unknown aspects of the problem prior to optimization.

In population-based non-Pareto approach, the population of an EA is used to diversify the search. This uses m subpopulations to optimize each objective independently. These subpopulations are then shuffled together to obtain a new population on which evolutionary operators are applied. Though it searches for multiple nondominated solutions concurrently in a single EA run, there is a possibility of a good compromise solution for all the objectives to be discarded as it is not the best in any of them. This approach has the main disadvantage of not directly incorporating the concept of Pareto dominance.

The third approach is to determine an entire Pareto optimal solution set or a representative subset. This method thereby attempts to promote the generation of multiple nondominated solutions by making use of the actual definition of Pareto-optimality. In real-life problems, Pareto optimal sets are often preferred since the final solution of the decision-maker is always a trade-off. Pareto optimal sets can be of varied sizes but the size of the Pareto set usually increases with the increase in the number of objectives.

1.5 Genetic Algorithm

Genetic algorithm (GA) is a stochastic search procedure first developed by Holland and is based on the mechanics of natural selection, genetics, and evolution (Goldberg 1989). Since GA simultaneously evaluates many points in the search space, it is likely to find a global solution of a given problem.

GA evolves a population of candidate solutions. Each solution that is called as an individual or chromosome represents a point in the search space. Each individual is decided by an evaluating mechanism to obtain its fitness value. Based on the fitness value and genetic operations, a new population is generated iteratively with each successive population referred to as a generation. GA uses three basic operators – selection, crossover, and mutation to manipulate the genetic composition of a population. Selection operators are used to select good individuals and form a mating pool to which the crossover operators will be applied. Crossover operators select individuals from the parental generation and interchange their genes to obtain new individuals. The individuals so obtained are subject to mutation. Mutation alters one or more gene values in an individual from its initial state resulting in new individuals helping to arrive at a better solution than before. This also prevents the population from stagnating at any local optima.

1.6 Multi-objective Genetic Algorithm

Being a population-based approach, GA is well-suited for solving a multi-objective problem. A generic single-objective GA can be easily modified to find a set of nondominated solutions in a single run. GA simultaneously searches different regions of a solution space to find a diverse set of solutions for difficult problems. The crossover operator of GA may exploit structures of good solutions with respect to different objectives to create new nondominated solutions in unexplored parts of the Pareto front. Some of the multi-objective evolutionary algorithms namely the weighted sum genetic algorithm, nondominated sorting genetic algorithm (NSGA-II), and multi-objective practical swarm optimization are described in the following sections.

1.6.1 Weighted Sum Genetic Algorithm (WSGA)

This approach of multi-objective optimization combines the individual objectives into a single composite function. The objectives to be optimized are scaled and randomly weighted. The weighted objectives are then summed up to obtain the fitness function. The fitness function can be implemented using Sum of Weighted Global Ratios (SWGR) method. This method is used because of its high performance and easy implementation as proved in literature (Bentley and Wakefield 1996). Based on SWGR, the weighted sum objective function for a given chromosome X is given in Eq. 1.4:

$$F(\mathrm{X}) = \sum_{j=1}^{k} w_j \frac{f_j(\mathrm{X}) - f_j^{\min}}{f_j^{\max} - f_j^{\min}} \tag{1.4}$$

where, $f_j(\mathrm{X})$ is the fitness of X with respect to the jth objective $f_j^{\max}$ and $f_j^{\min}$ are the maximal and minimal values for the jth objective, respectively. w_j is the weight associated with the jth objective. For this problem, a normalized weight vector is randomly generated for each solution (Murata and Ishibuchi 1995) so that the search space is widened in all directions instead of a fixed direction. The normalized weight vector for each objective is determined using Eq. 1.5:

$$w_j = \frac{\mathrm{rand}_j}{\sum_{i=1}^{k} \mathrm{rand}_i}, \quad i = 1, 2.........k \tag{1.5}$$

where, k is the number of objectives. The random number rand_j is a nonnegative real number so that the weights w_j generated lie in the closed interval [0, 1].

After evaluating the solutions, selection operator is applied to form a mating pool before genetic operators are applied. In binary tournament selection, the selection is done by making two randomly selected individuals to participate in the tournament and choosing the best among them. This scheme probabilistically duplicates some chromosomes and deletes others. However, better solutions have a higher probability of being duplicated in the next generation.

The crossover is applied at a higher probability P_{c} followed by mutation operation at a lesser probability P_{m} to generate a new population for the next generation. The algorithm is stopped when the required number of generations is reached. The generated output results in the optimal solution. The flow of the WSGA method is shown in Fig. 1.2.

1.6.2 Nondominated Sorting Genetic Algorithm II (NSGA-II)

The main objective of NSGA-II is to find multiple Pareto-optimal solutions in one single simulation run. As NSGA-II (Deb et al. 2002) works with a population of solutions, a simple multi-objective genetic algorithm (MOGA) can be extended to obtain a diverse set of solutions. Elitism is preserved by retention of good solutions that is achieved using suitable selection methods thereby enhancing convergence.

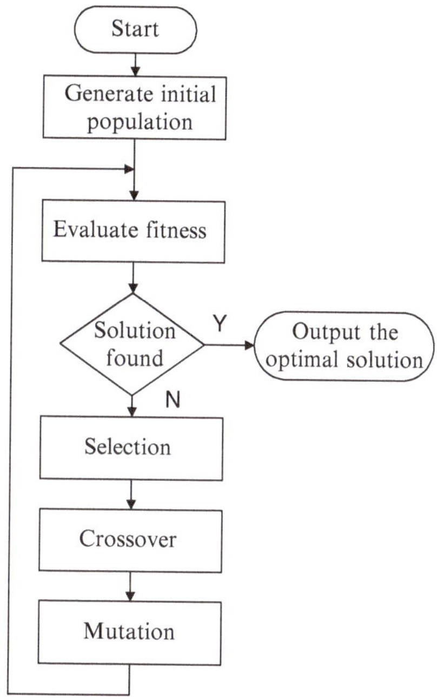

Fig. 1.2 Flowchart of WSGA

NSGA-II is an Elite Nondominated Genetic Algorithm (ENGA). The following are the main advantages of applying NSGA-II:

1. Employs nondominated sorting techniques to provide the solution as close to the Pareto-optimal solution as possible
2. Uses crowding distance technique to provide diversity in solution
3. Uses elitist techniques to preserve the best solution of current population in next generation

NSGA-II employs a fast nondominated sorting approach with a complexity of $O(mN^2)$ where m is the number of objectives and N is the population to be sorted. The flowchart of NSGA-II is shown in Fig. 1.3. An initial population X_t of size N is created. Two distinct entities are used in NSGA-II to validate the quality of a given solution. The first is to classify the population into various nondomination fronts using a nondominated sorting procedure. The second is to estimate the density of solutions surrounding a particular point in the population by means of crowding distance computation. After assigning a crowding distance value, the selection is carried out using a crowded comparison operator, referred to as crowded tournament selection, explained in the following section. Crossover and mutation operation is carried out on the selected solutions to generate the offspring population Y_t.

The offspring population Y_t is combined with the parent population X_t to form a new population Z_t of size $2N$. The population Z_t is sorted using nondominated sorting procedure. Since the previous population and current population members

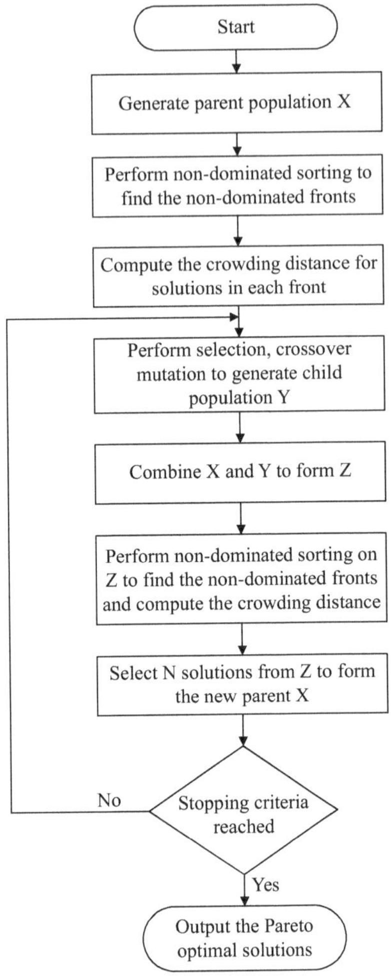

Fig. 1.3 NSGA-II algorithm

are included in Z_t, elitism is ensured. The solutions belonging to the first front F_1 are the best solutions and must be emphasized more than the other solutions. If the size of F_1 is smaller than N, then all the solutions of F_1 are placed in the new population. The remaining members of the population are chosen from subsequent nondominated fronts. Thus, solutions from the front F_2 are selected followed by solutions from F_3 and so on. This continues until a size of N is reached. When solutions from the last accepted front are considered, there may be more than the required number of solutions to form the new population. Hence, the solutions in this front are sorted according to crowding distance and solutions in the least crowded region are selected to fill all population slots. This new population X_{t+1} of size N is used in generating the new offspring population Y_{t+1}. The process repeats to generate the subsequent generations for obtaining the Pareto-optimal solutions.

1.6.2.1 Nondominated Sorting

Nondominated sorting is used to classify the population to identify the solutions for the next generation. The procedure for sorting is implemented as follows:

Step 1: For every solution p in population N.
Step 2: For every solution q in population N.
Step 3: If q not equal to p.

Compare p and q for all "m" objectives.

Step 4: If for any solution q, p is dominated by q, mark solution p as dominated.

The solutions that are not marked are called nondominated solutions, which form the first nondominated front in the population. The process is repeated with the remaining solutions for other higher nondomination fronts until the entire population is classified into u different fronts F_u.

1.6.2.2 Crowding Distance

Crowding distance calculation is the determination of Euclidian distance between each individual in a front based on their m objectives in the m dimensional space. Since the individuals are selected based on rank and crowding distance metric, every individual in the population is assigned a crowding distance value. The crowding distance is assigned front-wise. The crowding distance is calculated as follows:

Step 1: For all n individuals in each front F_u, initialize the distance to be zero. $F_u(cd_i) = 0$, where i corresponds to the ith individual in front F_u.
Step 2: For each objective function m, sort the individuals in front F_u based on objective m:

- $L = \text{sort}\ (F_u, m)$.
- Assign a large distance to the boundary points in L, $L(cd_1) = L(cd_n) = \infty$
- For the other points $i = 2$ to $(n - 1)$ in L is given by Eq. 1.6:

$$L(cd_i) = L(cd_i) + \frac{L_{i+1}.f_m - L_{i-1}.f_m}{f_m^{\max} - f_m^{\min}} \tag{1.6}$$

$L_i.f_m$ is the value of the mth objective function of the ith individual in L, $f_m^{\max}$ is the maximum value of mth objective, and $f_m^{\min}$ is the minimum value of mth objective.

1.6.2.3 Crowded Tournament Selection

Selection of individuals using crowded tournament selection operator is as given in the following.

A solution x is said to win the tournament with another solution y if any of the following conditions are true:

- If solution x has a better rank, that is, $r_x < r_y$
- If x and y hold the same rank and solution x has a better crowding distance than solution y, that is, $r_x = r_y$ and $cd_x > cd_y$

The first condition makes sure that the chosen solution lies on a better nondominated front. The second condition resolves the tie of both the solutions being on the same nondominated front by deciding on their crowded distance. The solution residing in the less crowded area wins, that is, it has a larger crowding distance.

1.6.3 *NSGA-II with Controlled Elitism (NSGA-II-CE)*

Elitism is an important issue to ensure diversity of individuals and get a better convergence. In NSGA-II, elitism is preserved at two places: once during the tournament selection and the other during elite preserving operation. However, NSGA-II does not assure the diversity where most of the members lie on the nondominated front of rank one and is a local Pareto front. In such cases, the elitist operation results in deletion of solutions belonging to nonelitist fronts. This slows down the search and result in premature convergence. Thus, an algorithm has to preserve diversity in both aspects – along and lateral to the Pareto-optimal front (Deb 2001) to ensure better convergence.

To achieve lateral diversity, the number of individuals in the current best nondominated front is limited adaptively. A predefined geometric distribution, as given by Eq. 1.7, is used to limit the number of individuals in each front:

$$N_i = rN_{i-1} \tag{1.7}$$

where N_i is maximum number of allowed individuals in the i-th front. The parameter $r(<1)$ is a user-defined parameter that specifies the reduction rate.

The main concept of the proposed method is to accommodate solutions from different nondominated levels to coexist in the population. Selection of solutions using this method for the next iteration is as follows. The combined 2N population of parent and offspring is sorted for nondomination. If the number of nondominated fronts in the 2N population is K, then according to the geometric distribution, the maximum number of individuals allowed in the ith front ($i = 1,2,\ldots,K$) in the new population of size N is given by Eq. 1.8. Since $r < 1$, the allowable number of individuals in front one is the maximum. Thereafter, each front is allowed to have an exponentially reducing number of solutions.

$$N_i = N\frac{1-r}{1-r^K}r^{i-1} \tag{1.8}$$

Equation 1.8 denotes the maximum number of individuals N_i in each front. If there does not exist exactly N_i individuals in such fronts, the following procedure is used to resolve the issue. The number of individuals in the first front N_1^t is counted initially. If $N_1^t > N_1$, then N_1 solutions are chosen using the crowded tournament selection. This selects N_1 solutions from the less crowded region. Otherwise, if $N_1^t \leq N_1$, all solutions are chosen and the number of remaining slots $X_1 = N_1 - N_1^t$ is counted. The maximum number of individuals in the second front is now increased to $N_2 + X_1$. Thereafter, the actual number of solutions in the second front is counted and compared with N_2 as earlier. This procedure is repeated till N individuals are selected. However, there could be some situations where the size of new population cannot reach N even after all $2N$ solutions are processed. The new population still has some space needed to be filled, especially when r is large. In such case, filling the population again continues with the remaining individuals from the first rank and continues to other ranks until the size of new population reaches N.

1.6.4 Hybrid NSGA-II with Pareto Hill Climbing (NSGA-II-PHC)

One promising strategy to improve the search ability of MOEAs is the hybridization of EAs with local search. Such EAs, which apply a separate local search process to refine individuals, are often referred to as memetic MOEAs. The memetic algorithms for multi-objective optimizations are dealt in Knowles and Corne (2004). The main issue in designing a local search procedure for the multi-objective case is to make it adaptable to optimize more than one objective simultaneously. Hill climbing is used as the local search method. The proposed hill-climbing method employs the concept of Pareto-dominance to obtain the improved set of solutions.

The proposed hybrid algorithm enhances the convergence rate by intensifying the search using a Pareto Hill Climbing (PHC) procedure and preserves diversity using the crowding-distance-based selection. The steps of the hybrid NSGA-II-PHC are as follows:

- A random initial population of size N is generated.
- The population is sorted into various nondomination levels using a nondominated sorting procedure and crowding distance is assigned to solutions at each level.
- Repeat till stopping criteria is reached.

- Apply selection, crossover, and mutation operators to the parent population to generate child population.
- Produce a population of size $2N$ by combining the parent with the child population together.
- Apply nondominated sorting method on the combined population and compute the crowding distance.
- Select N solutions from the combined population applying the controlled elitism selection procedure discussed in Sect. 1.6.3 to form the new parent.
- Apply PHC procedure to the new population using the procedure given in the following.

The PHC procedure begins by generating a neighborhood for the solution subject to local search using the mutation operator. Each neighbor is subject to dominance comparison and dominated neighbors are rejected. The process is repeated for defined number of iterations. The choice of solution subject to PHC is also crucial in reducing the computation time. The PHC procedure is described in the following.

For each front of the new population, repeat the following steps:

- Select X solutions from the less crowded area to apply PHC.
- Repeat the following steps for specified number of iterations for each selected solution:

  (a) Generate specified number of neighbors using mutation operation.
  (b) Compare the original solution with each of its neighbor for nondominance.
  (c) Replace it with the neighbor if the actual solution does not dominate the neighbor.
  (d) Repeat (b) and (c) until all neighbors are compared.

The number of solutions selected for applying PHC in a front X is equal to $2M + 1$, where M is the number of objectives considered. The reason for selecting $2M$ is because NSGA-II assigns infinite crowding distances to solutions having the maximum and/or minimum values on one or more objectives. Hence, $2M$ solutions may have infinite crowding distance at the maximum. Selecting $2M$ solutions for PHC leads to exploring only the edges of the Pareto region. To ensure exploration of at least one solution within the Pareto region, $2M + 1$ solutions are used.

1.7 Particle Swarm Optimization

Particle swarm optimization is one of the latest evolutionary optimization techniques developed by Eberhart and Kennedy (1995). PSO conducts searches using a population of particles. The particles which are potential solutions in the algorithm have a position represented by a position vector. The particles fly around in the multidimensional search space with the moving velocity of each particle represented by a velocity vector. The particles are evaluated for its fitness using a suitable function. Each particle keeps track of its own best position which is the best fitness it has achieved so far. Furthermore, the best position among all the particles in the population is kept track of. At each time step, by using the personal best position (*pbest*) and global best position (*gbest*), a new velocity for the particle is determined. Based on the updated velocities, each particle changes its position. Since all particles in PSO are kept as members of the population throughout the course of the searching process, PSO is the only evolutionary algorithm that does not implement survival of the fittest. These features enable PSO to maintain a balance between exploration and exploitation in the swarm and achieve fast convergence. The major advantage of PSO is easier implementation and requirement of few parameters adjustments.

1.8 Multi-objective Particle Swarm Optimization

The relative simplicity of PSO and the fact that it is a population-based technique have made it a natural candidate to be extended for multi-objective optimization. A comprehensive review of the various Multi-Objective Particle Swarm Optimization (MOPSO) algorithms is reported in literature (Margarita and Carlos 2006), which are considered to be promising areas of future research. New and sophisticated variants are introduced attempting to improve performance (Coello et al. 2004; Ho et al. 2005; Reyes-Sierra and Coello 2006).

The existing MOPSO are based on either an aggregating approach or a Pareto-based approach. The aggregating approach combines or aggregates all the objectives of the problem into a single one. The Pareto-based approach uses leader selection techniques based on Pareto dominance.

1.8.1 Weighted Sum Particle Swarm Optimization

The Weighted Sum PSO (WSPSO) described here uses a weighted aggregate approach to combine the objectives into a single objective. The fitness of the particles is evaluated using a function given by Eq. 1.4. The function uses the sum of weighted global ratios (SWGR) method. The weights are randomly generated to enrich searching directions and to obtain solutions with good diversity. The initial positions of the particles are set as the initial personal bests of the corresponding particles. The complete set of fitness values of all the personal best particles is examined and the position corresponding to the fittest value is used as the initial global best position (Abraham et al. 2008).

Three parameters contribute to the updation of the particle velocity: an internal contribution, proportional to its previous velocity; an exploratory contribution, proportional to the vicinity of the particle to its personal best position; and an exploitationary contribution, proportional to the closeness of the particle to the global best position.

The particle P_i adjusts its velocity V_{ij} through each dimension j by referring to pbest_{ij} and the swarm's best experience gbest_j using Eq. 1.9:

$$V_{ij} = wV_{ij} + C_1R_1(\text{pbest}_{ij} - P_{ij}) + C_2R_2(\text{gbest}_j - P_{ij}) \tag{1.9}$$

C_1 is the cognitive constant depicting the relative influence of the particle's personal best position on its velocity, C_2 is the social constant depicting the relative influence of the global best position on its velocity, and R_1 and R_2 are random numbers drawn from a uniform distribution U(0,1). The inertia weight, w, controls the momentum of the particle. A larger inertia weight moves toward global exploration while a smaller inertia weight helps in fine-tuning the current search space. The following weighting function is usually utilized:

$$w = w_{\text{ini}} - \frac{w_{\text{ini}} - w_{\text{fin}}}{\text{iter}_{\text{total}}} * \text{iter}_{\text{curr}} \tag{1.10}$$

where, w_{ini} represents the initial weight, w_{fin} – final weight, $\text{iter}_{\text{total}}$ – total number of iterations, and $\text{iter}_{\text{curr}}$ – current iteration number.

Once the velocities of all the particles are determined, they are used to obtain the updated position of the particle through each dimension, using Eq. 1.11:

$$P_{ij} = P_{ij} + V_{ij} \tag{1.11}$$

The WSPSO algorithm is given in the following:

Step 1: Initialization

- Generate N particles at random.
- Initialize the velocity of N particles to zero.

Step 2: Repeat until a stopping condition is reached

- Calculate the fitness of each particle
- Find the best vector *pbest* visited so far by each particle.
- Find the best vector *gbest* visited so far by the whole swarm.
- Update velocity and position of every particle using Eqs. 1.9 and 1.11.

1.8.2 Nondominated Sorting Particle Swarm Optimization (NSPSO)

To facilitate a multi-objective approach to PSO, a set of nondominated solutions must replace the single best individual as in the case of WSPSO. However, with the adaptation of Pareto-optimal concepts and the nondominated sorting process used in NSGA-II, PSO can be used to find the nondominated solutions in multi-objective optimization effectively. This modified algorithm is NSPSO (Li 2003). The standard PSO does not effectively utilize valuable nondomination comparisons while the personal best (pbest) of the particles are updated. NSPSO combines the *pbest* of *N* particles and the *N* particles offspring to form a population of 2*N* particles. These 2*N* particles are subject to nondomination comparisons for identifying different nondomination fronts. This provides the means for selecting the individuals in the better fronts, thereby providing the necessary selection pressure. Fitness values are assigned to individuals based on the front they belong to. Individuals that are completely nondominant in the entire population lie in the first front and given a rank value of 1. Individuals in the second front dominated by individuals in the front one are given a rank value of 2 and so on. To maintain population diversity, crowding distance assignment procedure is also adopted. The global best $gbest_n$ for the *nth* particle P_n is selected randomly from the top part of the first front having the highest crowding distance. Based on fitness value and crowding distance, *N* particles are selected as *pbest* particles. The selected *pbest* and *gbest* are used in updating the particles. The steps of basic NSPSO algorithm is shown in Fig. 1.4 and summarized in the following:

1. Generate an initial population of size *N*. The position and velocity of each particle in the swarm is initialized randomly within the specified limits. The personal best position of all the particles pbest_i, is set to P_i.
2. Sort the population based on nondomination and crowding distance ranking.
3. Assign each individual a rank value equal to its nondomination level.
4. Choose randomly, one individual as *gbest* for *N* times from the nondominated solutions of the best front.
5. Determine the new velocity and new position using the determined *gbest* and *pbest*.
6. Create a extended population of size 2*N* by combining the new position and their pbest.
7. Sort the extended population based on nondomination and crowding distance ranking.

1.8.3 Adaptive NSPSO (ANSPSO)

In addition to providing good convergence quality, the design of multi-objective optimization algorithms demands appropriate distribution quality of the determined Pareto optimal solutions in the whole objective space. Hence, the proposed adaptive NSPSO algorithm incorporates a dynamic setting of inertia weight *w* and the learning factors $C1$ *and* $C2$ for updating the particles.

1.8.3.1 Learning Factors

In the velocity update Eq. 1.9, higher values of C_1 ensure larger deviation of the particle in the search space, while the higher values of C_2 imply the convergence to the leader. To incorporate better compromise between the exploration and

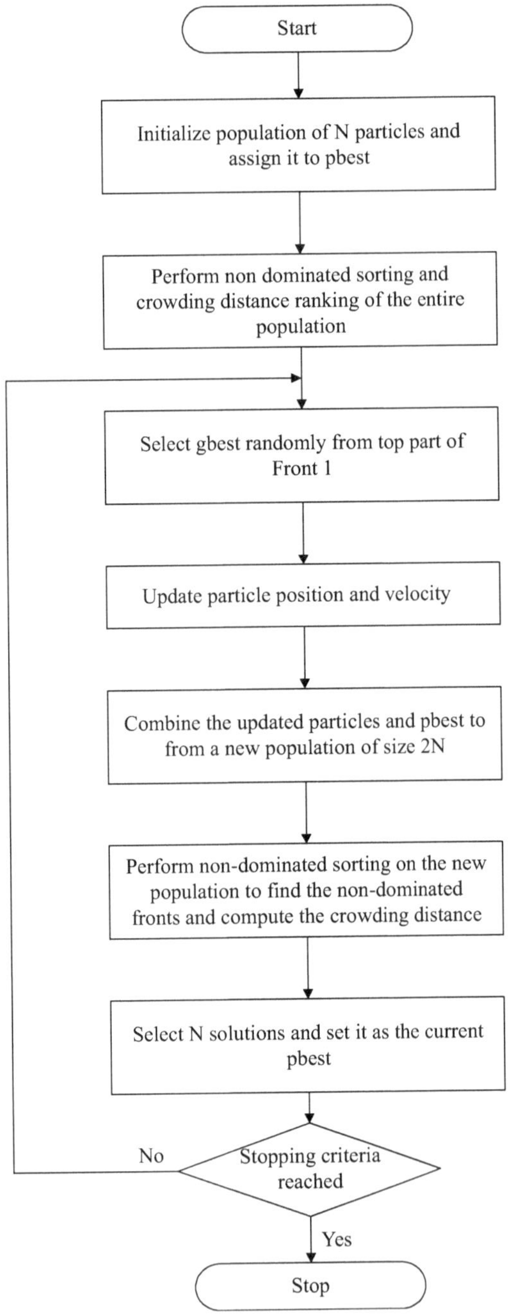

Fig. 1.4 NSPSO algorithm

exploitation of the search space in PSO, time-variant acceleration coefficients have been introduced in literature (Ratnaweera et al. 2004). The usage of these coefficients is exploited in the PSO. This aims to produce better Pareto optimal solutions. The coefficient C_1 is decreased from its initial value $C_1(i)$ to $C_1(f)$, whereas C_2 can be increased from $C_2(i)$ to $C_2(f)$. The values of C_1 and C_2 at iteration t are evaluated as follows:

$$C_1(t)=(C_1(f) - C_1(i)).\frac{t}{T}+C_1(i) \tag{1.12}$$

$$C_2(t)=(C_2(f) - C_2(i)).\frac{t}{T}+C_2(i) \tag{1.13}$$

where, T is the maximum number of iterations and t is the current iteration number.

1.8.3.2 Inertia Weight

The inertia weight (w) value in Eq. 1.9 plays a crucial role in the convergence quality of PSO algorithms. All the population-based search techniques rely on global exploration and local exploitation to achieve good performance. In the initial stages, more exploration is to be performed. This helps to gain more knowledge about the search space. In the later stages, the information gained so far is exploited for better results. Hence, the proposed method employs a crowding distance (Raquel and Naval 2005) value to calculate the inertia weight of each particle. The value of w is linearly decreased from an initial value ($w_{\max}$) to a final value ($w_{\min}$). The value of inertia weight at iteration t is obtained as follows:

$$w(t)=(w_{\max} - w_{\min}).\frac{t.e^{-cd}}{T}+w_{\min}.e^{-cd} \tag{1.14}$$

where T is the maximum number of iterations, t is the current iteration number and cd is the crowding distance. Particle's inertia weight tends to be $w_{\min}$ when $t=0$ and $w_{\max}$ when $t=T$ for smaller crowding distances. The inertia weight tends to be zero for the particle with larger crowding distance. This promotes diversity since a small crowding distance results in a high density of particles.

Equations 1.9 and 1.11 for updating the particles adapt the acceleration coefficients and inertia weight as given by Eqs. 1.12, 1.13, and 1.14. Hence, the parameters w, $C1$ and $C2$ used in updating the particles take different values for each particle in every iteration.

1.8.4 Hybrid NSPSO with Pareto Hill Climbing (NSPSO-PHC)

In this section, a hybrid optimization algorithm based on NSPSO is proposed. The flow chart of the proposed hybrid NSPSO is presented in Fig. 1.5. PSO-based

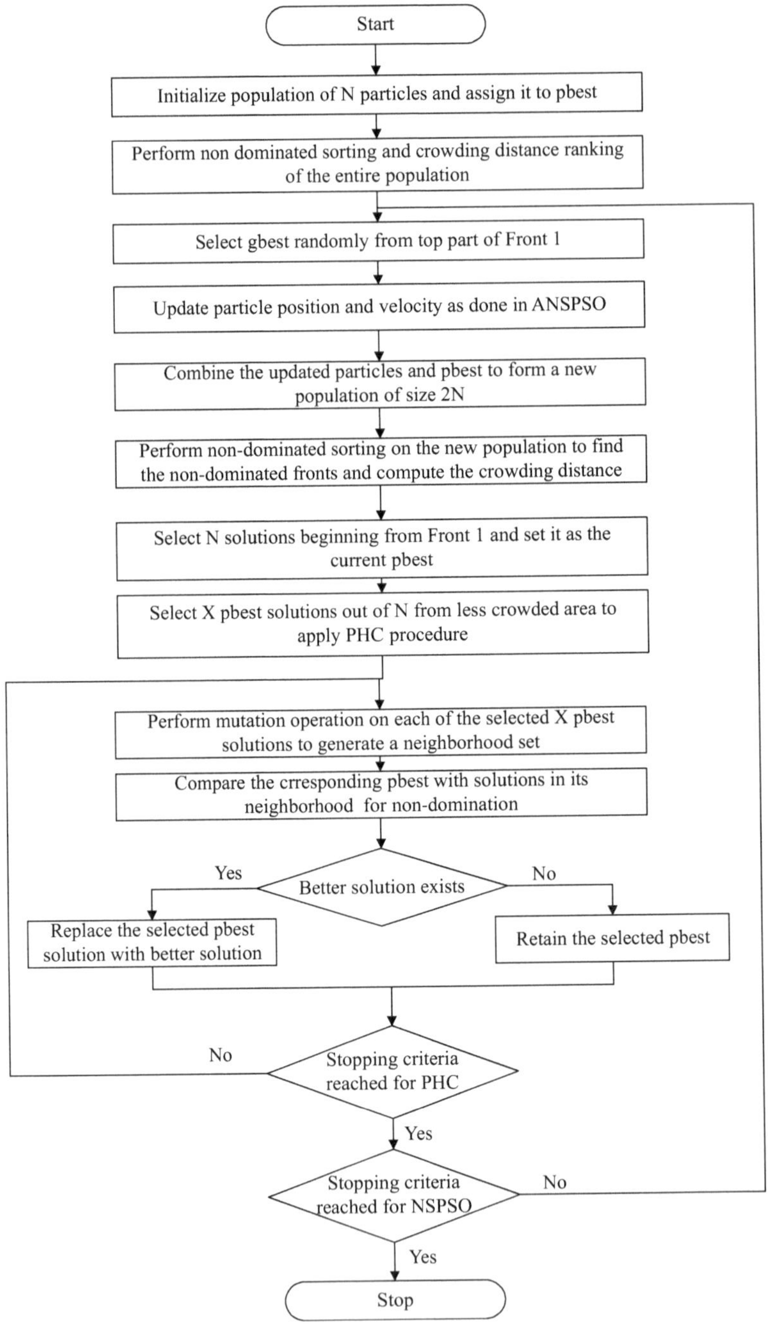

Fig. 1.5 NSPSO-PHC algorithm

mechanism is used for exploration and a multi-objective local search procedure is used for exploitation. To stress exploitation, a local search based on the PHC procedure is applied to good solutions with a specified probability. The balance of exploration and exploitation helps the hybrid NSPSO in obtaining good-quality solutions.

In PSO, the set of solutions in *pbest* is the optimal Pareto solution found during the searching procedure. The *pbest* solutions are employed to guide the evolution of the particles in the next iteration. In the hybrid NSPSO algorithm, PHC procedure is employed to a selected number of *pbest* solutions belonging to various fronts. The PHC procedure used employs a mutation operation to generate a neighborhood for each solution subject to PHC procedure. The selected solution is compared with all the solutions in the neighborhood set for nondomination. If a better solution exists in the neighborhood, the actual solution is replaced by a better solution to play the role of *pbest*. This is repeated for several iterations on each selected *pbest* particle. At the end of each iteration refined *pbest* particles are obtained, which will guide the particles in the next iteration. Thus, the local search procedure helps in better exploitation of the search space.

References

Abraham A, Liu H, Grosan C, Xhafa F (2008) Nature inspired meta-heuristics for grid scheduling: single and multi-objective optimization approaches, Studies in computational intelligence. Springer, Heidelberg, pp 247–272

Bentley PJ, Wakefield JP (1996) An analysis of multi-objective optimization within genetic algorithms. Technical Report ENGPJB96, University of Huddersfield, UK

Coello CAC (1999) A comprehensive survey of evolutionary-based multiobjective optimization techniques. Knowl Info Syst 1(3):129–156

Coello CAC, Pulido GT, Lechuga MS (2004) Handling multiple objectives with particle swarm optimization. IEEE Trans Evol Comput 8(3):256–279

Deb K (2001) Multi-objective optimization using evolutionary algorithms. Wiley, Chichester

Deb K, Pratab A, Agarwal S, Meyarivan T (2002) A fast and elitist multiobjective genetic algorithm: NSGA-II. IEEE Trans Evol Comput 6(2):182–197

Eberhart RC, Kennedy J (1995) A new optimizer using particle swarm theory. In: Proceedings of the sixth international symposium on micro machine and human science, pp. 39–43

Fonseca CM, Fleming PJ (1995) An overview of evolutionary algorithms in multiobjective optimization. Evol Comput 3(1):1–16

Goldberg DE (1989) Genetic algorithms in search of optimization and machine learning. Addison Wesley, Reading

Guliashki V, Toshev H, Korsemov C (2009) Survey of evolutionary algorithms used in multiobjective optimization. J Probl Eng Cybern Robot 60:42–54

Ho SL, Shiyou Y, Guangzheng N, Lo EWC, Wong HC (2005) A particle swarm optimization based method for multiobjective design optimizations. IEEE Trans Magn 41(5):1756–1759

Kaisa MM (1999) Nonlinear multiobjective optimization. Kluwer Academic Publishers, Boston

Knowles JD, Corne DW (2004) Memetic algorithms for multiobjective optimization: Issues, methods and prospects. In: Recent advances in memetic algorithms. Springer, Heidelberg

Li X (2003) A non-dominated sorting particle swarm optimizer for multi-objective optimization. In: Proceedings of genetic and evolutionary computation conference. Lecture notes in computer science, vol 2723. Springer, pp 37–48

Margarita RS, Carlos ACC (2006) Multi-objective particle swarm optimizers: a survey of the state-of-the-art. Int J Comput Intell Res 2(3):287–308

Marler RT, Arora JS (2004) Survey of multi-objective optimization methods for engineering. Struct Multidisc Optim 26:369–395

Murata T, Ishibuchi H (1995) MOGA: multi-objective genetic algorithms. IEEE Int Conf Evol Comput 1:289–294

Raquel CR, Naval PC Jr. (2005) An effective use of crowding distance in multiobjective particle swarm optimization. In: Proceedings of the genetic and evolutionary computation conference, Washington DC, USA, pp 257–264

Ratnaweera A, Halgamuge SK, Watson HC (2004) Self-organizing hierarchical particle swarm optimizer with time-varying acceleration coefficients. IEEE Trans Evol Comput 8(3):240–255

Reyes-Sierra M, Coello CAC (2006) Multi-objective particle swarm optimizers: a survey of the state-of-the-art. Int J Comput Intell Res 2(3):287–308

Zhou A, Qu BY, Li H, Zhao SZ, Suganthan PN, Zhang Q (2011) Multiobjective evolutionary algorithms: a survey of the state-of-the-art. J Swarm Evol Comput 1(1):32–49

Chapter 2
Hardware/Software Partitioning for Embedded Systems

M.C. Bhuvaneswari and M. Jagadeeswari

Abstract Current methods for designing embedded systems require specifying and designing hardware and software separately. Hardware/software partitioning is concerned with deciding which function is to be implemented in Hardware (HW) and Software (SW). This type of partitioning process is decided a priori to the design process and is adhered to as much as possible because any changes in this partition may necessitate extensive redesign. As partitioning is an NP hard problem, application of the exact methods tends to be quite slow for bigger dimensions of the problem. Heuristic/evolutionary methods have been proposed for partitioning problems. This chapter deals with multi-objective optimization of minimizing two objectives area and the execution time of the partition. To validate the efficiency of the algorithms, performance metrics are calculated. Experimental results for the HW/SW partition for mediabench and DSP benchmarks are tabulated and analyzed.

Keywords Hardware/software partitioning • Embedded systems • Pareto-optimal solutions • Performance metrics • Multi-objective optimization • Multi-objective evolutionary algorithms

2.1 Introduction

The complexity of very large scale system integration (VLSI) and embedded systems is increasing and the analysis and design of such complex systems is a nondeterministic polynomial time (NP) hard problem. Hardware/software co

---

M.C. Bhuvaneswari
Electrical and Electronics Engineering, PSG College of Technology, Coimbatore, India
e-mail: mcb@eee.psgtech.ac.in

M. Jagadeeswari (✉)
Electronics and Communication Engineering, Sri Ramakrishna Engineering College, Coimbatore, India
e-mail: jagadeeswari.m@srec.ac.in

M.C. Bhuvaneswari (ed.), *Application of Evolutionary Algorithms for Multi-objective Optimization in VLSI and Embedded Systems*, DOI 10.1007/978-81-322-1958-3_2,

design (HSCD) is the discipline of automating the design of complex embedded systems that function using both hardware and software. The central task of HSCD is hardware/software partitioning which is concerned with deciding which function is to be implemented in Hardware (HW) and Software (SW). This implementation aims at finding an optimal trade-off between conflicting parameters such as area and execution time. Deterministic methods for partitioning problems provide exact solutions, but when applied to complex problems they become computationally intractable. As partitioning is an NP hard problem, application of the exact methods tends to be quite slow for bigger dimensions of the problem. Heuristic/evolutionary methods have been proposed for partitioning problems (Schaumont 2013; Wu Jigang et al. 2010; Greg Stit 2008; He Jifeng et al. 2005; Zou Yi et al. 2004; Theerayod Wiangtong 2004; Arato et al. 2003; Dick and Jha 1997; Ernst et al. 1992). This chapter focuses on the application of multi-objective evolutionary algorithms for HW/SW partitioning of embedded systems.

This chapter mainly focuses on the algorithmic aspects of the central phase of HSCD, that is, functional partitioning. Functional partitioning divides a system functional specification into multiple subspecifications. Each subspecification represents the functionality of a system component such as HW or SW processor and is synthesized down to gates or compiled down to machine code. Functional partitioning results in both HW and SW implementations with less number of HW functional blocks. The embedded system to be partitioned is modeled as Directed Acyclic Graph (DAG). A DAG is a graph G (V, E), where V is the set of tasks (functional nodes) and the edge E represents the data dependency between the two tasks with $V \in \{v_1, v_2 \ldots v_i\}$ and $E \in \{e_1, e_2 \ldots e_i\}$. Each task is associated with five integer values: (1) HW-execution time (t_H), that is, the time required to execute the task on the HW module; (2) the HW implementation of the task requires area C_H on the HW module; (3) SW-execution time (t_S), that is, the time required to execute the task on the processor; (4) the SW implementation of the task requires memory C_S on the software module; and (5) the communication costs (C_C), which refers to the delay in transfer of data between HW and SW tasks. This cost is ignored if the transfer of data is between two nodes on the same side (i.e., two HW nodes or two SW nodes). Functional partitioning is done using evolutionary algorithms that decide the tasks to be implemented in HW or in SW. The evolutionary algorithms are iterated until the required number of generations is reached to obtain optimal HW/SW partitions.

In this chapter, three multi-objective optimization algorithms are illustrated: Weighted Sum Genetic Algorithm (WSGA) (Deb 2002), Nondominated Sorting Genetic Algorithm (NSGA-II) (Deb 2002), and Multi-objective Particle Swarm Optimization using Crowding Distance strategy (MOPSO-CD) (Tsou et al. 2006) applied for solving HW/SW partitioning problem. To validate the efficiency of the algorithms, performance metrics are calculated. Experimental results for the HW/SW partition for mediabench and DSP benchmarks are tabulated. The results obtained demonstrate that the NSGA-II is effective in solving HW/SW partitioning problem and provides good results for most of the benchmark circuits.

2.2 Prior Work on HW/SW Partitioning

Majority of the proposed partitioning algorithms are heuristic, due to the fact that the partitioning problem is NP-hard. The NP-hardness of the HW/SW partitioning was claimed by several researchers (Eles and Peng 1996; Kalavade and Lee 1994; Binh et al. 1996).

The proposed exact algorithms include branch-and-bound (D'Ambrosio and Hu 1994; Binh et al. 1996), dynamic programming (Madsen et al. 1997; Shrivastava and Kumar 2000), and integer linear programming (Arato et al. 2003). Two partitioning algorithms for HW/SW partitioning were presented by Arato et al. (2003): one based on Integer Linear Programming (ILP) and the other on Genetic Algorithm (GA). It was proved that ILP-based solution works efficiently for smaller graphs with several tens of nodes and yields optimal solutions, whereas GA gives near-optimal solutions on an average and works efficiently with graphs of hundreds of nodes. The performance of GA was found to be uniform, whereas wild oscillations were found in the run time of ILP.

Many researchers have applied general-purpose heuristics to HW/SW partitioning. Particularly, genetic algorithms have been extensively used (Wu Jigang et al. 2010; Greg Stit 2008; He Jifeng et al. 2005; Zou Yi et al. 2004; Theerayod Wiangtong 2004; Arato et al. 2003; Srinivasan et al. 1998; Ernst et al. 1992). Simulated Annealing (SA) was popularized by Kirkpatrick Vecchi (1983) as an alternative to greedy approaches, which are quickly trapped in local minima since they can only make downhill moves. This algorithm was also used by Theerayod Wiangtong (2004), Eles and Peng (1996), Ernst et al. (1992), and Lopez-Vallejo et al. (2000). Other less popular heuristics in this group are tabu search (Theeryod Wiangtong 2004; Eles and Peng 1996) and greedy algorithms (Chatha and Vemuri 2001). Some researchers used custom heuristics to solve HW/SW partitioning (Barros et al. 1993). This includes Global Criticality/Local Phase (GCLP) algorithm (Kalavade and Lee 1994), expert system of Lopez Vallejo et al. (2000), and binary constraint search algorithm by Vahid and Gajski (2001).

Concerning the system model, further distinctions can be made. In particular, many researchers consider scheduling as a part of partitioning (Kalavade and Lee 1994; Niemann and Marwedel 1997; Chatha and vemuri 2001; Lopez-Vallejo et al. 2000), whereas others do not (Eles and Peng 1996; Vahid et al. 1994). The problem of assigning communication events to link between HW and/or SW units was included by Dick and Jha (1997).

In a number of related articles, the target architecture consists of single SW and a single HW unit (Eles and Peng 1996; Gupta and De Micheli 1993; Henkel and Ernst 2001; Lopez-Vallejo and Lopez 2001; Vahid and Gajski 2001). Other researchers do not impose this limitation. Some researchers limit parallelism inside HW or SW (Vahid and Gajski 2001) or between HW and SW (Henkel and Ernst 2001). All of these algorithms aimed at optimization of only one objective. Jagadeeswari and Bhuvaneswari (2009) employed multi-objective optimization techniques for HW/SW partitioning of embedded systems.

2.3 Target Architecture

The characteristic of the target system architecture consists of one SW processor (aimed to execute C programs), one HW module (implements the VHDL/Verilog descriptions), and shared bus (communication between HW and SW) as shown in Fig. 2.1.

This target model serves as a platform onto which an HW/SW system is mapped. Both the HW and SW have their local memory and communicate with each other through the shared bus. There is no memory limitation in this model. The SW processor is a uniprocessing system and can execute only one task at a time, while the HW can execute multiple tasks concurrently.

2.4 Input Model

The embedded system to be partitioned is represented as a DAG. A DAG is a graph G (V, E), where V is the set of tasks (functional nodes) and E, the edge, represents the data dependency between the two tasks with $V \in \{v_1, v_2 \ldots v_i\}$ and $E \in \{e_1, e_2 \ldots e_i\}$ (Blickle 1996; Hidalgo and Lanchares 1997; Vahid 1997; Bakshi and Gajski 1999; Zou Yi et al. 2004). The example of such a DAG is shown in Fig. 2.2.

Each task (node) in the DAG is associated with SW area and SW time. The SW area (C_S) represents the SW memory utilized by the task and SW time (t_S) represents the execution time of the task if implemented in software processor. The HW implementation of each task is associated with an HW area and an HW time. The HW area (C_H) represents the area occupied by the task while implementing in HW and the HW time (t_H) is the execution time of the task if implemented in HW. If one of the two communicating tasks is implemented in HW, and the other in SW processor, then the communication cost (C_C) between them incurs a significant overhead and is considered during partitioning. If the two tasks are both in HW or both in SW, then the overhead is much lower and the communication cost is neglected. The communication cost in this context refers to the delay time required to transfer the data from HW module to SW module and vice versa.

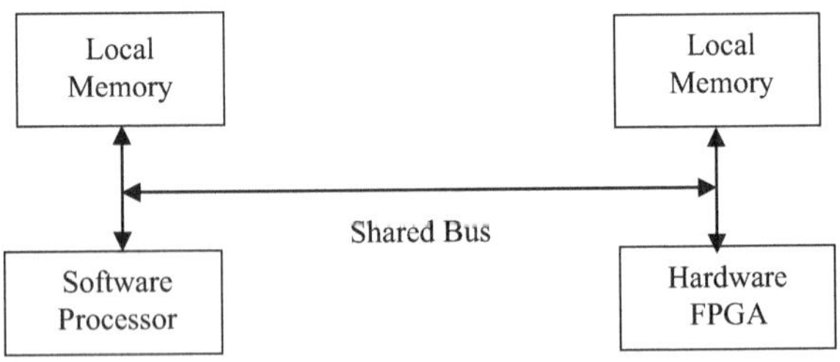

Fig. 2.1 Target architecture

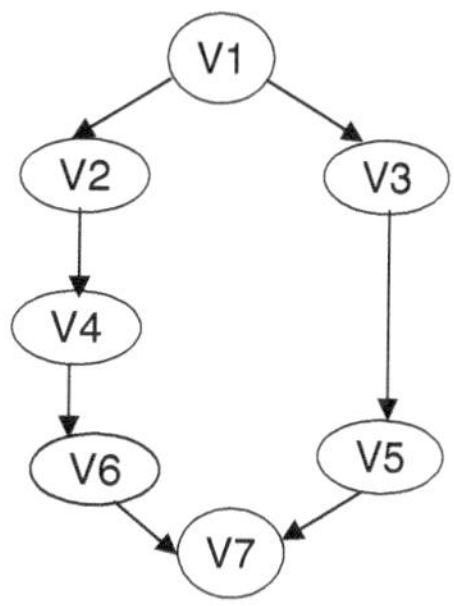

Fig. 2.2 Directed acyclic graph

2.5 Objective Function

During the HW/SW partitioning process, the evolutionary algorithms iterates to move each task mapped in SW to HW, which reduces the total execution time or tries to move task mapped in HW to SW, which reduces the total area. Multi-objective evolutionary algorithms can be employed for optimizing more than one parameter such as area, speed, power, and delay. In this work, the objectives used to guide the evolutionary algorithms through the optimization process are the total execution time (T) and total area (A), which is calculated using Eqs. 2.1 and 2.2, respectively. Power can be obtained as the sum of HW power and SW power. Optimization of power can be obtained by keeping area and time as constraints.

$$T = \sum_{\forall t_i \in HW} t_H + \sum_{\forall t_i \in SW} t_S + C_C \tag{2.1}$$

$$A = \sum_{\forall t_i \in HW} C_H + \sum_{\forall t_i \in SW} C_S \tag{2.2}$$

The total time is the sum of the execution times of the tasks implemented in the HW (t_H) and SW (t_S) and the delay in communication (C_C) between the HW and SW tasks and vice versa. The total area is the sum of the silicon area of the tasks implemented in the HW (C_H) and SW memory utilized (C_S) by the tasks when implemented in the SW (Hou and Wolf 1996; Zou Yi et al. 2004).

2.6 Encoding Procedure

The chromosomes (individuals) are made up of units called genes. Group of individuals form a population. The individuals are characterized by a chromosome with an amount of genes equal to the number of functional nodes with each gene representing a task in the system. Binary encoding scheme is employed. The chromosome is characterized as an individual defined by $\{t_1, t_2, t_3, t_4 \ldots t_n\}$,

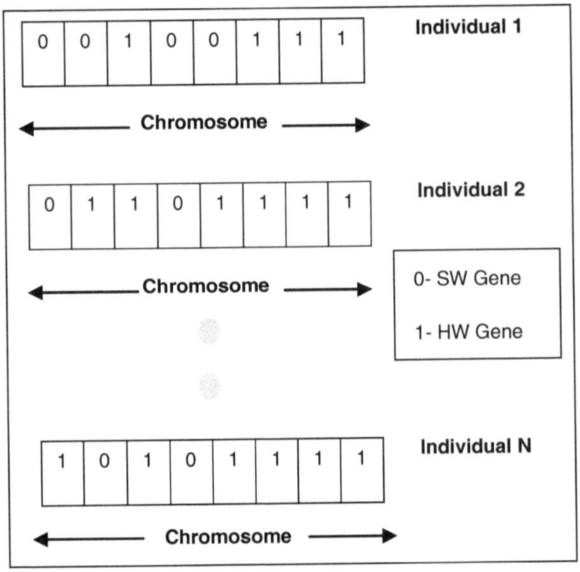

Fig. 2.3 Sample population for HW/SW partitioning problem

t_i € {1, 0}, i € {1, 2. . .n}, where "n" is the number of tasks in the embedded system. If $t_i = 1$, the corresponding block is implemented in the HW and if $t_i = 0$, then the corresponding block is implemented in the SW as shown in Fig. 2.3.

2.7 Performance Metric Evaluation

When different algorithms exist for solving a particular problem, it becomes necessary to compare these in terms of their performance. There are two distinct goals in multi-objective optimization: (1) discover solutions as close to the pareto-optimal solutions as possible and (2) find solutions as diverse as possible in the obtained nondominated front. These two goals are orthogonal to each other as the first one requires a search toward pareto-optimal region while the second goal needs a search along the pareto-optimal front.

2.7.1 Metrics Evaluating Closeness to True Pareto-Optimal Front

These metrics explicitly compute a measure of the closeness of a set of M solutions obtained from the known true pareto-optimal set (T) (Deb 2002).

2.7.1.1 Error Ratio (ER)

The metric ER simply counts the number of solutions that are not the members of the true pareto-optimal set T. It is calculated using Eq. 2.3.

$$ER = \frac{\sum_{i=1}^{M} E_i}{M} \tag{2.3}$$

M is the total number of solutions obtained. The value of $E_i = 1$ if the solution "i" does not belong to the true pareto-optimal solutions and $E_i = 0$ otherwise. The smaller the value of ER, the better is the convergence to true pareto-optimal front. The metric takes a value between zero and one. An $ER = 0$ means all the solutions are members of T and $ER = 1$ means no solution is a member of T.

2.7.1.2 Generational Distance (GD)

This metric determines the average distance of the obtained solutions from the set T. GD is obtained using Eq. 2.4:

$$GD = \frac{\left(\sum_{i=1}^{M} d_i^T\right)^{1/T}}{M} \tag{2.4}$$

where d_i is the Euclidean distance between the solution $i \in M$ and the nearest member of T, calculated using Eq. 2.5:

$$d_i = \min_{j=1}^{|T|} \sqrt{\sum_{n=1}^{N} \left(f_n^{(i)} - f_n^{(j)}\right)^2} \tag{2.5}$$

where $f_n^{(j)}$ is the nth objective function value of the jth member of the true pareto-optimal solutions (T) and N is the total number of objectives used in the problem. An algorithm that generates a small value of GD implies that better solutions are obtained.

2.7.1.3 Maximum Pareto-Optimal Front Error (MFE)

This metric computes the worst distance d_i among all the members of the obtained solutions M, that is, the maximum distance among the distances d_i is calculated using Eq. 2.5. A lesser value of this metric indicates that the obtained solutions are closer to set T.

2.7.2 *Metrics Evaluating Diversity among Nondominated Solutions*

These metrics find the diversity among the obtained nondominated solutions (Deb 2002).

2.7.2.1 Spacing (S)

This metric calculates the relative distance measure between the two consecutive solutions, using Eq. 2.6:

$$S = \sqrt{\frac{1}{M}\sum_{i=1}^{M}\left(d_i - \overline{d}\right)^2} \tag{2.6}$$

where d_i is calculated using Eq. 2.5 and $\overline{d}$ is the mean value of the Euclidean distance measured as given by Eq. 2.7:

$$\overline{d} = \sum\nolimits_{i=1}^{M} \frac{d_i}{M} \tag{2.7}$$

A value of zero for the metric spacing indicates that all members of the obtained solutions are equidistantly spaced.

2.7.2.2 Spread (Δ)

This metric measures the spread of the obtained solutions and is calculated using the formula given in Eq. 2.8:

$$\Delta = \frac{\sum\nolimits_{n=1}^{N} d_n^{\mathrm{es}} + \sum\nolimits_{i=1}^{M} \left|d_i - \overline{d}\right|}{\sum\nolimits_{n=1}^{N} d_n^{\mathrm{es}} + M\overline{d}} \tag{2.8}$$

where d_i and $\overline{d}$ are calculated using Eqs. 2.5 and 2.7, respectively. The parameter d_n^{es} is the distance between the extreme solutions corresponding to the nth objective function. This metric takes a value of zero for an ideal distribution only when $d^{\mathrm{es}} = 0$ and all d_i values are identical to their mean $\overline{d}$. $d^{\mathrm{es}} = 0$ means that only true pareto-optimal solutions exist in the obtained solutions and if d_i values are identical to their mean $\overline{d}$, then the distribution of intermediate solutions obtained is uniform. Such a solution set is the ideal outcome of any multi-objective evolutionary algorithm. For an ideal distribution of solutions, the value of $\boldsymbol{\Delta}$ is zero. In another case, the distribution of the obtained solution is uniform but they are clustered in one place, then the distribution will make all $|d_i - \overline{d}|$ values zero and will cause nonzero values for d_n^{es}. The quantity $\mathbf{\Delta}$ lies within [0, 1]. For bad distributions, the Δ values can be more than one as well.

2.7.2.3 Weighted Metric (W)

One of the convergence metrics and one of the diversity preserving metrics are combined using Eq. 2.9 to form the weighted metric W:

$$W = A.GD + B.\Delta \tag{2.9}$$

With $A + B = 1$. In this metric, the GD, which specifies the converging ability, and Δ, which specifies the diversity-preserving ability of the algorithm are combined. GD takes a small value for good convergence and Δ takes a small value for good diversity-preserving algorithm. The algorithm with an overall small value of W means that the algorithm is efficient in both the aspects. To combine the two metrics, the weights A and B are chosen depending on the importance of the performance metric. To give equal weightage to the two metrics, the values are chosen as A = B = 0.5. The performance metrics namely ER, GD, MFE can be determined only when the true pareto-optimal solutions are known for the specified problem. The other metrics S, Δ, and W can be determined even when the true pareto-optimal solutions are unknown.

2.8 Experimental Results

The algorithm is programmed in C language and was run using 2.80 GHz, Pentium-IV processor with 1 GB RAM. The algorithms are iterated using the parameter values tabulated in Table 2.1 for the 10-node DAG (Zou Yi et al. 2004). The true pareto-optimal solutions for this system are determined using exhaustive search so that the performance metrics ER, MFE, and GD can be calculated. The adaptation of WSGA, NSGA-II, and MOPSO-CD algorithms for multi-objective optimization of HW/SW partitioning problem for 10-node DAG is obtained and analyzed for minimizing two objectives: area (A) and task execution time (T) (Jagadeeswari and Bhuvaneswari 2009). The algorithms are tested on a 10-node DAG with the input values indicated in Tables 2.1 and 2.2 (Zou Yi et al. 2004). Table 2.1 shows the data available on each node and Table 2.2 shows the delay in communication between each node. The parameter settings used in the algorithms are given in Table 2.3.

Tables 2.4, 2.5, and 2.6 list all the pareto-optimal solutions obtained using the three algorithms WSGA, NSGA-II, and MOPSO-CD, respectively along with the HW/SW implementation for 10-node system. From the results, it is found that both

Table 2.1 Data table for 10-node system

Node no.	SW area (KB)	SW time (μs)	HW area (KB)	HW time (μs)
V1	24	526	166	318
V2	45	209	158	133
V3	58	511	201	255
V4	23	366	152	242
V5	80	21	351	6
V6	55	178	194	92
V7	24	92	162	57
V8	237	204	2342	8
V9	67	277	239	117
V10	59	317	261	122

Table 2.2 Data exchange of 10-node system

	V1	V2	V3	V4	V5	V6	V7	V8	V9	V10
V1	–	5	20	19	0	0	3	16	0	14
V2	0	0	0	7	14	18	0	13	0	2
V3	0	0	0	0	0	0	0	4	19	0
V4	0	0	0	0	0	0	0	0	0	7
V5	0	0	0	0	0	5	0	0	0	0
V6	0	0	0	0	0	0	0	0	11	0
V7	0	0	0	0	0	0	0	0	10	0
V8	0	0	0	0	0	0	0	0	0	19
V9	0	0	0	0	0	0	0	0	0	13
V10	0	0	0	0	0	0	0	0	0	0

Table 2.3 Parameter settings used in multi-objective algorithms for 10-node DAG

S. No.	Parameter	WSGA	NSGA-II	MOPSO-CD
1.	Population size (2N)	20	20	–
2.	Crossover probability	0.8	0.8	–
3.	Mutation probability	0.02	–	–
4.	Type of crossover	Two point	Two point	–
5.	Number of generations	100	30	100

NSGA-II and MOPSO-CD are able to determine all the pareto-optimal solutions. MOPSO-CD determines the output with less search time than NSGA-II and WSGA for the 10-node DAG. To compare the performance of the three algorithms and to determine the performance metrics, true pareto-optimal solutions for the 10-node DAG are obtained using exhaustive search method. It is carried out by calculating the nondomination output for all the objectives in 2^{10} binary combinations. The plot of such a search is shown in Fig. 2.4.

The quality of the results obtained using WSGA, NSGA-II, and MOPSO-CD is measured using the performance metrics that identify how close and diverse are the obtained pareto-optimal solutions with the true pareto-optimal solutions. The

Table 2.4 Optimal solutions obtained using WSGA

Output	Node implementation	Total area (KB)	Total execution time (μs)	Number of process in HW
Solution 1	1111111111	4,226	1,556	10
Solution 2	1111011111	3,955	1,596	9
Solution 3	1111010111	3,817	1,701	8
Solution 4	1111001111	3,816	1,798	8
Solution 5	1111111011	2,121	1,812	9
Solution 6	1111011011	1,850	1,852	8
Solution 7	1111010011	1,712	1,957	7
Solution 8	1111001011	1,711	2,054	7
Solution 9	1110011001	1,549	2,097	6
Solution 10	1111001010	1,509	2,144	6
Solution 11	1110010001	1,411	2,182	5
Solution 12	1111000010	1,371	2,249	5
Solution 13	1110001000	1,208	2,327	4
Solution 14	1000010001	1,155	2,380	3
Solution 15	1110000000	1,070	2,412	3
Solution 16	0000010001	1,013	2,539	2
Solution 17	1000000000	814	2,570	1
Solution 18	0000000000	672	2,701	0

Table 2.5 Optimal solutions obtained using NSGA-II

Output	Node implementation	Total area (KB)	Total execution Time (μs)	Number of Process in HW
Solution 1	1111110111	4,088	1,406	9
Solution 2	1011110011	1,870	1,407	7
Solution 3	1011010011	1,599	1,419	6
Solution 4	1011000011	1,460	1,463	5
Solution 5	1011110011	1,331	1,606	7
Solution 6	1011000010	1,258	1,671	7
Solution 7	1010000010	1,129	1,800	3
Solution 8	1011000000	1,086	1,933	3
Solution 9	0010000010	987	1,971	2
Solution 10	0011000000	944	2,142	2
Solution 11	0010000000	815	2,233	1
Solution 12	1000000000	814	2,570	1
Solution 14	0001000000	801	2,610	1
Solution 15	0000000000	672	2,701	0

different performance metrics such as Error Ratio (ER), Generational Distance (GD) or Standard Deviation (γ), Maximum Pareto-Optimal Front Error (MFE), Spacing (S), Spread (Δ), and Weighted Metric (W) for the 10-node DAG using WSGA, NSGA-II, and MOPSO-CD are calculated using the formulas mentioned

Table 2.6 Optimal solutions obtained using MOPSO-CD

Output	Node implementation	Total area (KB)	Total execution time (μs)	Number of process in HW
Solution 1	1111110111	4,088	1,406	9
Solution 2	1011110011	1,870	1,407	7
Solution 3	1011010011	1,599	1,419	6
Solution 4	1011000011	1,460	1,463	5
Solution 5	1011110011	1,331	1,606	7
Solution 6	1011000010	1,258	1,671	7
Solution 7	1010000010	1,129	1,800	3
Solution 8	1011000000	1,086	1,933	3
Solution 9	0010000010	987	1,971	2
Solution 10	0011000000	944	2,142	2
Solution 11	0010000000	815	2,233	1
Solution 12	1000000000	814	2,570	1
Solution 14	0001000000	801	2,610	1
Solution 15	0000000000	672	2,701	0

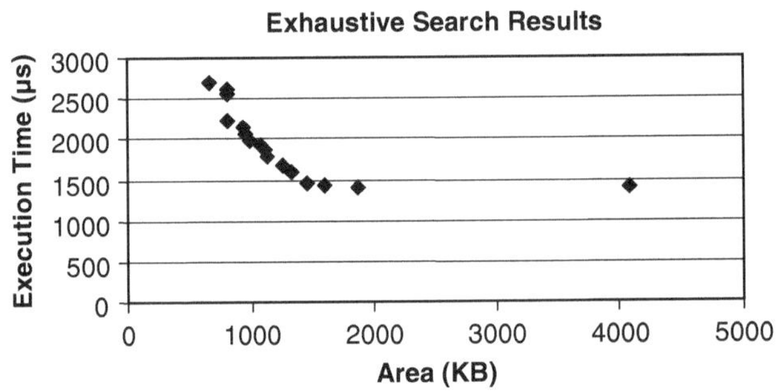

Fig. 2.4 True pareto-optimal solutions

Table 2.7 Comparison of performance metrics for 10-node system

Method	ER	GD(γ)	S	MFE	Δ	W	Run time (s)
MOPSO-CD	**0.0**	**0.0**	**74.9**	**0.0**	**0.5**	**0.20**	**0.34**
NSGA-II	**0.0**	**0.0**	**74.9**	**0.0**	**0.5**	**0.20**	1.05
WSGA	0.22	34.55	102.69	93.60	3.84	20.19	9.62

and tabulated in Table 2.7. The first three metrics are calculated based on the true pareto-optimal solutions obtained with the exhaustive search method. The other three metrics can be determined without the knowledge of true pareto-optimal solutions. It is noted that the performance metrics obtained by using NSGA-II and MOPSO-CD are the same, since both the algorithms find the same pareto-optimal solutions. The performance of MOPSO-CD algorithm with respect to run time is better than NSGA-II and WSGA for the 10-node DAG.

Table 2.8 Simulation results for benchmark DAG

Bench mark DAG	Node/ edges	Algorithm used	Mean area (KB)	Mean execution time (μs)	Run time (s)
HAL FILTER	11/8	GA	381.88	154.53	0.17
		WSGA	379.24	135.23	8.33
		NSGA-II	354.50	124.74	1.42
		MOPSO-CD	**323.70**	**124.74**	**0.16**
MPEG MOTION VECTOR	32/29	GA	979.29	543.67	1.48
		WSGA	889.54	423.59	12.55
		NSGA-II	849.61	398.70	2.81
		MOPSO-CD	**840.33**	**343.22**	**0.59**
JPEG DOWN SAMPLE	49/52	GA	1032.79	888.53	5.11
		WSGA	1015.25	723.85	15.68
		NSGA-II	**885.14**	**459.43**	4.06
		MOPSO-CD	965.11	601.22	**3.77**
MPEG	114/164	GA	4551.35	1579.56	20.64
		WSGA	3971.25	1497.47	34.23
		NSGA-II	**3312.50**	**981.50**	11.52
		MOPSO-CD	3523.50	1281.50	**10.71**
JPEG IDCT	120/162	GA	4339.88	3543.50	21.26
		WSGA	3455.58	2563.98	37.09
		NSGA-II	**3010.31**	**1969.46**	**11.23**
		MOPSO-CD	3145.38	2368.69	11.66

An effective HW/SW partition for the mediabench and DSP benchmark DAGs was obtained using WSGA, NSGA-II, and MOPSO-CD multi-objective optimization algorithms and compared with single-objective GA. All the algorithms are run for 100 generations. The results are analyzed and tabulated in Table 2.8. It is observed that MOPSO-CD algorithm performed better than NSGA-II, WSGA, and GA for HAL FILTER and MPEG MOTION VECTOR benchmarks with less mean area and mean task execution time. The run time of MOPSO-CD algorithm is less than GA, WSGA, and NSGA-II.

For all other benchmark DAGs applied, NSGA-II performs better than MOPSO-CD, WSGA, and GA with lesser mean area and mean task execution time. The run time of MOPSO-CD is comparable to NSGA-II and less when compared to GA and WSGA. From the experimental results it is observed that MOPSO-CD performs better for DAGs with lesser nodes (less than 50 nodes) and NSGA-II is found to perform better than MOPSO-CD, GA and WSGA for DAGs with larger number of nodes (more than 50 nodes).

2.9 Summary

In this chapter, three multi-objective evolutionary algorithms, WSGA, NSGA-II and MOPSO-CD, are dealt for HW/SW partitioning of embedded systems. Multioptimal solutions for the functional partitioning of embedded systems using

WSGA, NSGA-II, and MOPSO-CD was obtained. The pareto-optimal solutions obtained for a 10-node DAG are analyzed. The closeness to true pareto-optimal solutions and the diversity among the pareto-optimal solutions found is determined by calculating the performance metrics. It is seen that both NSGA-II and MOPSO-CD algorithms performed better than WSGA for 10-node DAG. The reason may be that crowding distance strategy retains the optimal solutions obtained in each generation until the final generation is reached. The algorithm was tested on several mediabench and DSP benchmark DAGs. For the mediabench and DSP benchmark applications, MOPSO-CD was found to perform better for graphs less than 50 nodes. NSGA-II was found to perform better compared to GA, WSGA, and MOPSO-CD, for nodes greater than 50 and HW/SW implementations with lesser mean area and mean execution time are obtained.

References

Arato P, Juhasz S, Mann ZA, Orban A, Papp D (2003) Hardware/software partitioning in embedded system design. In: Proceedings of the IEEE international symposium on intelligent signal processing, 4–6 Sept 2003, Budapest, Hungary, pp 197–202

Bakshi S, Gajski D (1999) Partitioning and pipelining for performance-constrained hardware/software system. IEEE Trans Very Large Scale Integr Syst 7(4):419–432

Barros E, Rosenstiel W, Xiong X (1993) Hardware/software partitioning with UNITY. In: Proceedings of 2nd international workshop on hardware-software codesign- CODES/CASHE '93, Austria, 24–27 May 1993, pp 210–217

Binh NN, Imai M, Shiomi A, Hikichi N (1996) A hardware/software partitioning algorithm for designing pipelined ASIPs with least gate counts. In: Proceedings of the 33rd annual conference on design automation, 3–7 June 1996, Las Vegas, pp 527–532

Blickle T (1996) Theory of evolutionary algorithms and applications to system synthesis. Dissertation, Swiss Federal Institute of Technology, Zurich

Chatha KS, Vemuri R (2001) MAGELLAN: Multiway hardware-software partitioning and scheduling for latency minimization of hierarchical control-dataflow task graphs. In: Proceedings of CODES'01, 25–27 Apr 2001, Copenhagen, pp 42–47

D'Ambrosio JD, Hu X (1994) Configuration-level hardware/software partitioning for real-time embedded systems. In: Proceedings of third international workshop on hardware/software codesign, 22–24 Sept 1994, Grenoble, pp 34–41

Deb K (2002) Multi-objective optimization using evolutionary algorithms. Wiley, Chichester

Dick RP, Jha NK (1997) MOGAC: a multi-objective genetic algorithm for the co-synthesis of hardware-software embedded systems. In: Proceedings of IEEE/ACM international conference on computer-aided design, 9–13 Nov 1997, San Jose, pp 522–529

Eles P, Peng Z (1996) System level hardware/software partitioning based on simulated annealing and Tabu Search. Des Autom Embed Syst 2(1):5–32

Ernst J, Henkel J, Benner T (1992) Hardware software co synthesis for microcontrollers. IEEE Des Test Comput 10(4):64–75

Greg Stit (2008) Hardware/software partitioning with multi-version implementation exploration. In: Proceedings of 18th ACM Great Lakes symposium on VLSI, 04–06 May 2008, USA, pp 143–146

Gupta RK, De Micheli G (1993) Hardware-software co-synthesis for digital systems. IEEE Des Test Comput 10(3):29–41

He Jifeng, Huang, Dang, Pu Geguang, Qiu, Zong Yan Yi, Wang (2005) Exploring optimal solution to hardware/software partitioning for synchronous model. J Formal Aspects Comput 17(4):587–611

Henkel J, Ernst R (2001) An approach to automated hardware/software partitioning using a flexible granularity that is driven by high-level estimation techniques. IEEE Trans VLSI Syst 9(2):273–289

Hidalgo JI, Lanchares J (1997) Functional partitioning for hardware-software codesign using genetic algorithms. In: Proceedings of 23rd EUROMICRO conference, 01–04 Sept 1997, Budapest, Hungary, pp 631–638

Hou J, Wolf W (1996) Process partitioning for distributed embedded systems. In: Proceedings of fourth international workshop on hardware/software co-design, Mar 1996, Pittsburgh, Pennsylvania, pp 70–76

Jagadeeswari M, Bhuvaneswari MC (2009) An efficient multi-objective genetic algorithm for hardware-software partitioning in embedded system design: ENGA. Int J Comput Appl Technol 36(3/4):181–190

Kalavade A, Lee E (1994) A global criticality/local phase driven algorithm for the constrained hardware/software partitioning problem. In: Proceedings of third international workshop on hardware/software co design, 22–24 Sept 1994, Grenoble, pp 42–48

Kirkpatrick S Jr, Gelatt CD, Vecchi MP (1983) Optimization by simulated annealing. Comput Sci J 220(4598):671–680

Lopez Vallejo M, Grajal J, Lopez JC (2000) Constraint-driven system partitioning. In: Proceedings of design automation and test in Europe, Jan 2000, Paris, France, pp 411–416

Lopez Vallejo M, Lopez JC (2001) Multi-way clustering techniques for system level partitioning. In: Proceedings of the 14th IEEE ASIC/SOC conference, 12–15 Sept 2011, Arlington, pp 242–247

Madsen J, Grode J, Knudsen P (1997) Hardware/software partitioning using the LYCOS system. Hardware/software codesign: principles and practices, Springer, US

Niemann R, Marwedel P (1997) An algorithm for hardware/software partitioning using mixed linear programming. Des Autom Embed Syst 2(2):165–193

Schaumont P (2013) A practical introduction to hardware/software codesign, 2nd edn. Springer Science+Business media, New York

Shrivastava A, Kumar M (2000) Optimal hardware/software partitioning for concurrent specification using dynamic programming. In: Proceedings of thirteenth international conference on VLSI design, 03–07 Jan 2000, Calcutta, India, pp 110–113

Srinivasan V, Radhakrishnan S, Vemuri R (1998) Hardware software partitioning with integrated hardware design space exploration. In: Proceedings of design automation and test in Europe, Feb 1998, Paris, France, pp 28–35

Theerayod Wiangtong (2004) Hardware/software partitioning and scheduling for reconfigurable systems. Dissertation, Imperial College, London

Tsou CS, Fang HH, Chang HH, Kao CH (2006) An improved particle swarm pareto optimizer with local search and clustering, vol 4247, Lecture notes in computer science. Springer-Verlag, Berlin, Heidelberg, pp 400–406

Vahid F (1997) Modifying min-cut for hardware and software functional partitioning. In: Proceedings of fifth international workshop on hardware/software co-design, 24–26 Mar 1997, Germany, p 43

Vahid F, Gajski DD (2001) Incremental hardware estimation during hardware/software functional partitioning. IEEE Trans VLSI Syst 3:516–521

Vahid F, Jie Gong, Daniel D Gajski (1994) A binary constraint search algorithm for minimizing hardware during hardware/software partitioning. In: Proceedings of the conference on European design automation EURO-DAC '94, Grenoble, pp 214–219

Wu Jigang, Qiqiang Sun, Thambipillai Srikanthan (2010) Multiple-choice hardware/software partitioning: computing models and algorithms. In: Proceedings of international conference on computer engineering and technology, April 2010, China, pp 61–65

Zou Yi, Zhenquan Zhuang, Huanhuan Chen (2004) HW/SW partitioning based on genetic algorithm. In: Proceedings of the congress on evolutionary computation, 19–23 June 2004, pp 628–633

Chapter 3
Circuit Partitioning for VLSI Layout

M.C. Bhuvaneswari and M. Jagadeeswari

Abstract Partitioning is a technique to divide a circuit or system into a collection of smaller parts (components). Circuit partitioning problem is a well-known NP hard problem and requires efficient heuristic algorithms to solve it. The problem involves dividing the circuit net list into two subsets. The balanced constraint is an important constraint that obtains an area-balanced layout without compromising the min-cut objective. The number of edges belonging to two different partitions is the cut-cost of a partition. This chapter deals with multi-objective genetic algorithms for the optimization of two objectives: area imbalance and cut cost. The objective is to separate the cells into two partitions so that the number of interconnections between the partitions can be minimized and the cells are evenly distributed across the layout surface. MCNC benchmark circuits are used to validate the performance of the multi-objective evolutionary algorithms.

Keywords Circuit partitioning • Multi-objective optimization • WSGA • NSGA-II • MOPSO • Pareto-optimal solutions

3.1 Introduction

In Integrated Circuit (IC) design, the physical design is a step in the standard design cycle, which follows after the circuit design and transforms a circuit description into the physical layout that describes the position of the cells and routes the

---

M.C. Bhuvaneswari
Electrical and Electronics Engineering, PSG College of Technology, Coimbatore, India
e-mail: mcb@eee.psgtech.ac.in

M. Jagadeeswari (✉)
Electronics and Communication Engineering, Sri Ramakrishna Engineering College, Coimbatore, India
e-mail: jagadeeswari.m@srec.ac.in

M.C. Bhuvaneswari (ed.), *Application of Evolutionary Algorithms for Multi-objective Optimization in VLSI and Embedded Systems*, DOI 10.1007/978-81-322-1958-3_3,

interconnections between them. As the technology scales down, it is possible to place a large number of logic gates on a single chip. Despite this fact, it may become necessary to partition a circuit into several subcircuits and implementation of those subcircuits as ICs. The reason may be that the circuits are too large to be placed on a single chip or because of input/output pin limitations. The following are the steps to be performed during the physical design process. (1) Partitioning: The circuit has to be subdivided to generate macro cells. (2) Floor planning: The partitioned cells have to be placed on the layout surface. (3) Placement: Algorithms are employed to generate optimal positioning of the cells. (4) Routing: This process is carried out using global and detailed routing. (5) Layout compaction: The process is done to optimize the overall area of the layout surface. Therefore, the physical design process becomes an NP-hard problem and requires efficient heuristic algorithms to solve it.

The physical design process has strong serial interdependencies between the substeps. Partitioning is the first step in the physical design cycle and forms the base for obtaining the optimized design. In deep-submicron design, partitioning defines local and global interconnects and has significant impact on circuit performance. This chapter deals with circuit partitioning as it acts as a facilitator of divide-and-conquer meta-heuristics for floor planning, timing, and placement optimization, and is also an effective way to solve problems of high complexity. The major objectives of partitioning are to obtain minimum number of interconnections between partitions and to achieve balanced partitioning, given bound for area of each partition. Therefore, a good partitioning method can significantly improve circuit performance and reduce layout costs.

The problem of circuit partitioning is encountered frequently in VLSI design, for example, in circuit layout, circuit packaging, and circuit simulation. The objects to be partitioned in VLSI design are logic gates or instances of standard cells. The objective is to separate the cells into two blocks. A balance constraint is imposed to ensure that the partitioned blocks contain equal area. Placement algorithms perform partitioning repeatedly in an attempt to find a placement that minimizes overall routing. After the circuit is partitioned into blocks, each block can be assigned a location on the layout surface. By minimizing the number of interconnections between the blocks, the algorithm is able to minimize the number of nets that must be routed over long distances.

In this chapter, three multi-objective optimization algorithms, Weighted-Sum Genetic algorithm (WSGA) (Deb 2002), Nondominated Sorting Genetic Algorithm (NSGA-II) (Deb 2002), and Multi-objective Particle Swarm Optimization using Crowding Distance strategy (MOPSO-CD) (Tsou et al. 2006), are applied to partitioning problems for placement and routing in the design of VLSI layout circuits and compared. The objects to be partitioned are logic gates or cells. The circuit-partitioning problem is represented in graph theoretic notation as weighted graph with the gates represented as nodes and the wires connecting them as edges. The weights of the nodes represent the sizes of the corresponding gates and the weights on the edges represent the number of wires connecting the gates. The objective is to separate the cells into two partitions so that the number of interconnections between the partitions can be minimized and the cells are evenly distributed across the layout surface. The experimental results for MCNC benchmark circuits are evaluated.

3.2 Prior Work on Circuit Partitioning

In 1970, Kernighan and Lin (KL) introduced the first "good" graph bisection heuristic (Ouyang et al. 2002). KL iteratively swaps the pair of unlocked modules with the highest gain (Gajski et al. 1994; Gerez 1999). Fiduccia and Mattheyses (FM) (1982) presented a KL-inspired algorithm that reduced the time per pass to linear in the size of the net list. Circuit partitioning using tabu search and genetic algorithm was done by Areibi and Vannelli (1993), and Sait et al. (2003, 2006).

For the min-cut bisection problem, Johnson et al. (1989) conducted an extensive empirical study of simulated annealing versus iterative improvement approaches, using various random graphs as a test bed. Tabu search was proposed by Glover (1989) as a general combinatorial optimization technique. The tabu list can be viewed as an alternative to the locking mechanism in KL and FM. Mardhana and Ikeguchi (2003) proposed a neuro-search-based method for solving a VLSI net-list partitioning problem. They explained the key concepts of neuro-search methods to support a VLSI net-list partitioning program (Al-Abaji 2002). Genetic partitioning methodology for placement of VLSI circuits was done by Sipakoulis et al. (1999) and Srinivasan et al. (2001).

In the mid-1990s, Kennedy and Eberhart (1995, 1997) enunciated an alternative solution to the complex nonlinear optimization problem by emulating the collective behavior of bird flocks, particles, and sociocognition (Kennedy et al. 2001) and called their brainchild Particle Swarm Optimization (PSO) (Carlisle and Dozier 2000; Shi and Eberhart 1998; Bergh and Engelbrecht 2002; Brandstatter and Baumgartner 2002; Clarc and Kennedy 2002; Salman et al. 2002; Wang et al. 2003; Hassan et al. 2004; Sudholt and Witt 2008). All of these algorithms aimed at optimization of a single objective.

Multi-objective optimization was carried out by different researchers. Multi-objective circuit partitioning for cut size and path delay minimization was proposed by Cristinel Ababei et al. (2002). Fuzzy rules were incorporated into single-objective GA, Tabu search and Simulated evolution, in order to handle the multi-objective cost function designed for delay and power by Khan et al. (2002).

3.3 Illustration of Circuit Bipartitioning Problem

It is computationally a hard problem to partition a set of circuit elements into two or more subsets so that connectivity between the nodes across the blocks can be minimized. In the worst case, it takes exponential time to divide a set of circuit

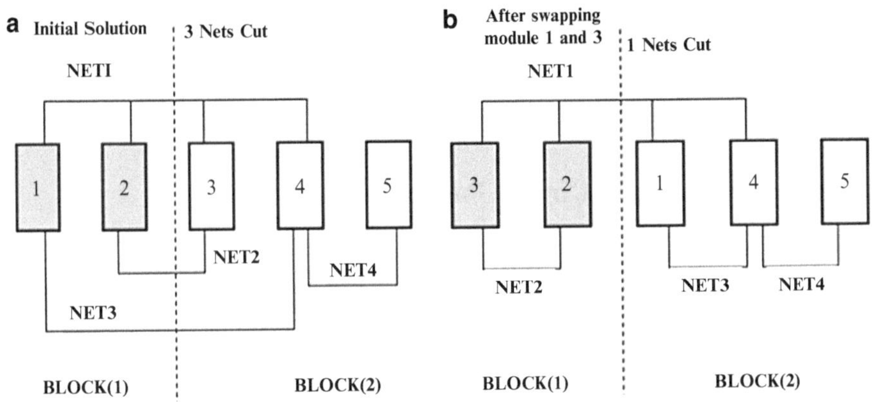

Fig. 3.1 Illustration of circuit bipartition

elements into "k" blocks by enumerating all possible permutations in which "n" circuit components can be divided into "k" equal blocks of size $p = n/k$. The total number of unique ways of partitioning the graph is given in Eq. 3.1. The equation becomes more complex when the number of nodes n is large and the bipartitioning problem ($k = 2$) leads to nonpolynomial time (Mazumder and Rudnick 1999).

$$N_k = \frac{n!}{k!(p!)^k} \tag{3.1}$$

The circuit bipartitioning problem can be formally represented in graph theoretic notation as a weighted graph with the components represented as nodes and the wires connecting them as edges. The weights of the nodes represent the area of the corresponding components and the weights of the edges represent the number of wires connecting the components. In its general form, the partitioning problem consists of dividing the nodes of the graph into two disjoint subsets so that the sum of weights of the nodes in each subset remain the same (minimize area imbalance) and the sum of weights of the edges connecting nodes in different subsets is minimized. The weights on the edges are assumed a value of one (Mazumder and Rudnick 1999).

Consider the circuit with five nodes and four nets as in Fig. 3.1a. During bipartitioning, the five nodes may be separated between two blocks, that is, nodes 1 and 2 may be placed in block 1 and nodes 3, 4, and 5 in block 2. The net cut obtained is three. The nodes 1 and 3 are swapped from one block to another to minimize the objective. After swapping, the numbers of signal nets that interconnect the components between the blocks is minimized to one as shown in Fig. 3.1b.

3.4 Circuit Bipartitioning Using Multi-objective Optimization Algorithms

Multi-objective optimization for circuit bipartitioning is carried out using proper encoding and selection of fitness function.

3.4.1 *Encoding Procedure*

The GA starts with several alternative solutions to the optimization problem, which are considered as individuals in the population. These solutions are encoded as binary strings called chromosomes. Figure 3.2 shows the encoding scheme to represent the bipartitioning problem. Nodes B1, B2, B4, and B8 are located in partition 1 and hence the gene values are fixed as zero. Nodes B3, B5, B6, and B7 have gene value 1 as they belong to partition 2 (Areibi and Vannelli 1993). The initial population is constructed at random and the individuals are evaluated using the fitness function. The populations of the individuals are iterated until the number of generations is reached to achieve the Pareto-optimal solutions.

3.4.2 *Fitness Function Formulation*

The partitioning algorithm divides the set of cells (nodes) into two disjoint subsets, so that the sum of cell areas in each subset is balanced and the sum of the costs (C) of nets connected to cells in the two different subsets is minimized. The two objectives used to guide the evolutionary algorithms in the optimization process are given in Eq. 3.2 (Mazumder and Rudnick 1999) and Eq. 3.3 (Al-Abaji 2002; Drechsler et al. 2003):

$$C = \sum_{i=1}^{m} \sum_{j=1}^{n} C_{ij} \text{ is minimized} \tag{3.2}$$

and

$$\sum_{i=1}^{m} M_i = \sum_{j=1}^{n} N_j \tag{3.3}$$

where C is the total cost of the cut, C_{ij} is the cost of an edge connecting nodes in partition i and partition j, and m and n are the number of nodes in partition 1 and partition 2, respectively. M_i and N_j are the total area of the nodes in partition 1 and partition 2, respectively. The objective is to reduce the value of the cut cost

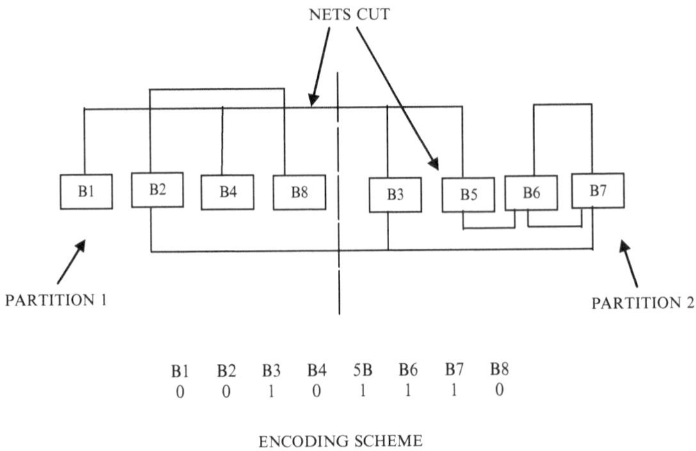

Fig. 3.2 Chromosome representation for circuit bipartitioning problem

(*C*) so that the number of nets cut is optimized and also to balance the two partitions by placing equal number of components in each partition. This is evaluated by calculating the area imbalance. Area imbalance is calculated as the sum of squares of the deviation from this ideal value and both overfilled and underfilled partitions are penalized. Area imbalance is small for small deviations and much larger for large deviations.

3.5 Experimental Results

The multi-objective algorithms are programmed in C language in Linux environment and is run using 2.80 GHz, Pentium-IV processor with 1GB RAM. The performance of multi-objective algorithms WSGA, NSGA-II, and MOPSO-CD is tested using MCNC benchmark circuits (Jagadeeswari and Bhuvaneswari 2010). The parameter settings used in the multi-objective algorithms are tabulated in Table 3.1.

Table 3.2 shows the benchmark circuit details of MCNC benchmark circuits used for circuit bipartitioning. The experimental results for circuit bipartitioning obtained using WSGA, NSGA-II, and MOPSO-CD are tabulated in Table 3.3. The mean values of cut cost and area imbalance are obtained for the average of 10 runs. It is noted that the mean cut cost and mean area imbalance of the bipartition obtained using NSGA-II and MOPSO-CD are less for Fract benchmark. For other larger benchmark circuits, NSGA-II obtains lesser mean cut cost and mean area imbalance than MOPSO-CD and WSGA.

The results tabulated in Table 3.3 show that the mean cut-cost and the area imbalance obtained by NSGA-II are consistently better than WSGA and MOPSO-CD for almost all the MCNC benchmark circuits. The mean cut cost obtained using

Table 3.1 Parameter settings used in evolutionary algorithms for circuit bipartitioning

S. No.	Parameter	GA, WSGA	NSGA-II	MOPSO-CD
1.	Population size	100	100	–
2.	Crossover probability	1	1	–
3.	Mutation probability	0.02	–	–
4.	Type of crossover	Two point	Two point	–
5.	Number of generations	100	50	100

Table 3.2 MCNC benchmark circuit details

Benchmark circuit	Total number of nodes	Total number of nets
Fract	149	164
Primary 1 (Prim 1)	833	905
Struct	1,952	1,920
Primary 2 (Prim 2)	3,014	3,030
Biomed	6,417	5,711
Industry 1 (Ind 1)	3,085	2,595

Table 3.3 Simulation results of GA, WSGA, NSGA-II, and MOPSO-CD for MCNC benchmark circuits

	GA		WSGA		NSGA-II		MOPSO-CD	
Bench mark Circuit	Mean cut cost	Mean area imbalance	Mean cut cost	Mean area imbalance	Mean cut cost	Mean area imbalance	Mean cut cost	Mean area imbalance
Fract	15.1	12.5	14.2	8.5	**9.8**	**2.2**	**9.8**	4.5
Prim 1	95.0	18.0	95.0	16.7	**85.4**	**5.8**	95.0	12.4
Struct	170.0	27.3	165.0	24.5	**128.2**	**12.8**	160.2	23.6
Prim 2	200.5	39.2	191.3	31.0	**163.6**	**15.4**	190.3	18.6
Biomed	163.2	29.5	153.2	20.5	**105.5**	**11.2**	155.2	18.2
Ind 1	69.0	13.2	67.5	10.2	**35.3**	**8.2**	67.5	10.0

multi-objective algorithms for MCNC benchmarks are better than the mean cut cost obtained using conventional GA (Areibi and Vannelli 1993; Mazumder and Rudnick 1999). The performance of the multi-objective algorithms is obtained by calculating the percentage improvement of NSGA-II over WSGA and MOPSO-CD and tabulated in Table 3.4.

From the comparison results, it is observed that NSGA-II outperforms the other two algorithms with respect to two objectives. This shows that NSGA-II obtains smaller area imbalance and fewer net cuts compared to WSGA and MOPSO-CD for MCNC benchmarks in circuit bipartitioning applications.

The run time comparison for the three algorithms for MCNC benchmark circuits is tabulated in Table 3.5. It is seen that for most of the benchmark circuits NSGA-II is faster in yielding better bipartition solutions compared to WSGA and MOPSO-CD. Figure 3.3 shows the run time comparison plot for MCNC benchmark circuits.

Table 3.4 Comparison of performance of evolutionary algorithms for circuit bipartitioning

	Percentage improvement of NSGA-II over GA		Percentage improvement of NSGA-II over WSGA		Percentage improvement of NSGA-II over MOPSO-CD	
Bench mark circuit	Mean cut	Mean area imbalance	Mean cut	Mean area imbalance	Mean cut	Mean area imbalance
Fract	35.09	82.4	30.99	74.12	0	51.11
Prim 1	10.10	67.78	10.11	65.27	10.11	53.23
Struct	24.59	53.11	22.30	47.76	19.98	45.76
Prim 2	18.40	60.71	14.48	50.32	14.03	17.20
Biomed	35.27	62.03	31.14	45.37	32.02	38.46
Ind 1	48.84	37.88	47.70	19.61	47.70	18.00
Average	**28.72**	**61.15**	**26.12**	**50.41**	**20.64**	**37.29**

Table 3.5 Run time comparison of evolutionary algorithms for circuit bipartitioning

	Run time (seconds)			
Benchmark circuit	GA	WSGA	NSGA-II	MOPSO-CD
Fract	15.20	10.27	5.24	**4.32**
Prim 1	65.55	52.55	**12.36**	31.43
Struct	70.25	65.32	**26.55**	44.54
Prim 2	80.95	72.25	**36.77**	50.23
Biomed	75.32	68.54	**33.45**	49.54
Ind 1	69.50	60.35	**25.24**	39.12

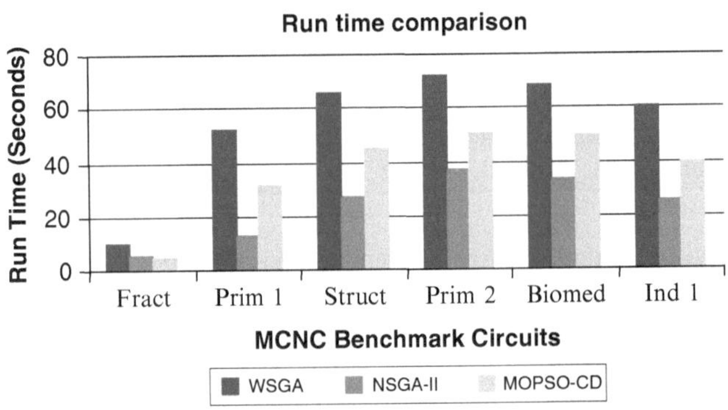

Fig. 3.3 Performance comparison of multi-objective algorithms for MCNC benchmark circuits

3.6 Summary

The multi-objective algorithms, namely, WSGA, NSGA-II and MOPSO-CD, are applied to VLSI circuit bipartitioning problem for placement and routing in layout applications. These algorithms are tested on MCNC benchmark circuits. For most of the MCNC benchmark circuits applied, NSGA-II yields better bipartitions, which contain 26% fewer cut nets, 50% less area imbalance than WSGA, and 21% fewer cut nets, 37% less area imbalance than MOPSO-CD. NSGA-II obtains the Pareto-optimal solutions faster than WSGA and MOPSO-CD.

References

Al-Abaji RH (2002) Evolutionary techniques for multi-objective VLSI net list partitioning. Dissertation, King Fahd University of Petroleum and Minerals, Dhahran, Kingdom of Saudi Arabia

Ababei C, Selvakumaran N, Bazargan K, Karypis G (2002) Multi objective circuit partitioning for cut size and path-based delay minimization. In: Proceedings of the 2002 IEEE/ACM international conference on computer-aided design, 10–14 Nov, Sanjose, USA, pp 181–185

Areibi S, Vannelli A (1993) A combined eigenvector tabu search approach for circuit partitioning. In: Proceedings of IEEE custom integrated circuits conference, San Diego, 09–12 May 1993, pp 9.7.1–9.7.4

Bergh F, Engelbrecht A (2002) A new locally convergent particle swarm optimizer. In: Proceedings of conference on systems, man and cybernetics, Hammamet-Tunisia, pp 96–101

Brandstatter B, Baumgartner U (2002) Particle swarm optimization – mass-spring system analog. IEEE Trans Magn 38:997–1002

Carlisle A, Dozier G (2000) Adaptive particle swarm optimization to dynamic environment. In: Proceedings of international conference on artificial intelligence, Mexico, pp 429–443

Clarc M, Kennedy J (2002) The particle swarm – explosion, stability, and convergence in a multidimensional complex space. IEEE Trans Evol Comput 6(1):58–73

Deb K (2002) Multi-objective optimization using evolutionary algorithms. John Wiley & Sons, USA

Drechsler R, Gunther W, Eschbach T, Linhard L, Angst G (2003) Recursive bi-partitioning of net lists for large number of partitions. J Syst Architect 49(12–15):521–528

Fiduccia CM, Mattheyses RM (1982) A linear time heuristic for improving network partitions. In: Proceedings of nineteenth design automation conference, IEEE Press, Piscataway, pp 175–181

Gajski DD, Vahid F, Narayau S, Gong J (1994) Specification and design of embedded system. Prentice Hall, NJ, USA

Gerez SH (1999) Algorithm for VLSI design automation. John Wiley & Sons, USA

Glover F (1989) Tabu search – Part I. ORSA. J Comput 1(3):190–206

Hassan R, Cohanim B, de Weck O (2004) A comparison of particle swarm optimization and the genetic algorithm. American Institute of Aeronautics and Astronautics, pp 1–13

Jagadeeswari M, Bhuvaneswari MC (2010) Application of evolutionary algorithms for partitioning in VLSI and Embedded Systems. Dissertation, Anna University, Chennai, 2010

Johnson DS, Aragon CR, McGeoch LA, Schevon C (1989) Optimization by simulated annealing: an experimental evaluation, part I, graph partitioning. Operat Res 37(6):865–892

Kennedy J, Eberhart RC (1995) Particle swarm optimization. In: Proceedings of IEEE international conference on neural networks IV, July 1995, Piscataway, pp 1942–1948

Kennedy J, Eberhart, R (1997) A discrete binary version of the particle swarm algorithm. In: Proceedings of international conference on systems man and cybernetics, 12–15 Oct 1997, Orlando, pp 4104–4109

Kennedy J, Eberhart RC, Shi Y (2001) Swarm intelligence. Morgan Kaufmann Publishers, San Francisco

Kernighan BW, Lin S (1970) An efficient heuristic procedure for partitioning graphs. Bell Syst Techn J 49(2):291–307

Khan JA, Sait SM, Minhas MR (2002) Fuzzy bias less simulated evolution for multi-objective VLSI placement. In: IEEE CEC 2002, Hawaii, USA, 12–17 May 2002, pp 1642–1647
Mardhana E, Ikeguchi T (2003) Neuro search: a program library for neural network driven search meta-heuristics. In: Proceedings of 2003 international symposium on circuits and systems, 25–28 May 2003, Bangkok, Thailand, pp V-697–V-700
Mazumder P, Rudnick EM (1999) Genetic algorithms for VLSI design layout and test automation. Prentice Hall, New York
Ouyang M, Toulouse M, Thulasiraman K, Glover F, Deogun JS (2002) Multilevel cooperative search for the circuit/hyper graph partitioning problem. IEEE Trans Comput-Aided Des Integr Circuits Syst 21(6):685–693
Sait SM, El-Maleh AH, Al-Abaji RH (2003) General iterative heuristics for VLSI multi-objective partitioning. In Proceedings of the 2003 international symposium on circuits and systems, ISCAS'03, 25–28 May 2003, Bangkok, Thailand, pp V-497–V-500
Sait SM, El-Maleh AH, Al-Abaji RH (2006) Evolutionary algorithms for VLSI multi-objective net list partitioning. Int J Eng Appl Artif Intell, 19(3):257–268
Salman A, Ahmad I, Al-Madani S (2002) Particle swarm optimization for task assignment problem. Microprocess Microsyst 26:363–371
Shi Y, Eberhart R (1998) Parameter selection in particle swarm optimization. In: Proceedings of 7th annual conference evolutionary programming, 25–27 Mar 1998, San Deigo, pp 591–600
Sipakoulis GC, Karafyllidis I, Thanailakis A (1999) Genetic partitioning and placement for VLSI circuits. In Proceedings of sixth IEEE international conference on electronics, circuits and systems ICECS'99, 05–08 Sept 1999, Pafos, Cyprus, pp 1647–1650
Srinivasan V, Govindarajan S, Vemuri R (2001) Fine-grained and coarse-grained behavioral partitioning with effective utilization of memory and design space exploration for multi-FPGA architectures. IEEE Trans VLSI Syst 9(1):140–158
Sudholt D, Witt C (2008) Run time analysis of binary PSO. In: Proceedings of the 10th annual conference on genetic and evolutionary computation, 16 July 2008, New York, pp 135–142
Tsou CS, Fang HH, Chang HH, Kao CH (2006) An improved particle swarm pareto optimizer with local search and clustering. In: Proceedings of sixth international conference SEAL, China, 15–18 Oct 2006, pp 400–406
Wang K, Huang L, Zhou C, Pang W (2003) Particle swarm optimization for traveling salesman problem. In: Proceedings of the second international conference on machine learning and cybermetics, 24–27 Aug 2003, Xi-an, China, pp 1583–1585

Chapter 4
Design of Operational Amplifier

M.C. Bhuvaneswari and M. Shanthi

Abstract The integrated circuit (IC) market growth has been explosive over the last decade. The design of complex ICs has been possible due to technology improvements and design automation tools. The design automation of analog integrated circuit is inevitable considering recently emerging discipline such as System on Chip (SOC) design. The design of the analog portion of a mixed-signal integrated circuit often requires large fraction of the overall design time, despite the fact that the analog circuit is often relatively a small portion of the overall circuit. This brings about the absolute necessity to develop Computer Aided Design (CAD) tools to automate the analog circuit synthesis process. There is a need for high-performance analog electronic circuit satisfying large number of specifications. An achievement of fast, low-power, and low-area integrated circuits is the major requirement in the electronic industry.

The Operational Amplifier (OpAmp) is the most versatile and widely used building blocks in analog electronics. The CMOS OpAmp design problem is a complex and tedious task, which requires many compromises to be made between conflicting objectives. The performance of an OpAmp is characterized by number of parameters, such as, gain, power, slew rate, and area. These performance parameters are determined by transistor dimensions, bias current, and compensation capacitance values. The synthesis of CMOS OpAmp is translated to a multi-objective optimization task. Evolutionary algorithm is applied for the OpAmp design because of the multi-objective nature of the design problem and large search space. In this chapter, synthesis of CMOS OpAmp circuit using Weighted Sum

M.C. Bhuvaneswari
Electrical and Electronics Engineering, PSG College of Technology, Coimbatore, India
e-mail: mcb@eee.psgtech.ac.in

M. Shanthi (✉)
Electronics and Communication Engineering, Kumaraguru College of Technology,
Coimbatore, India
e-mail: shanthi.m.ece@kct.ac.in

M.C. Bhuvaneswari (ed.), *Application of Evolutionary Algorithms for Multi-objective Optimization in VLSI and Embedded Systems*, DOI 10.1007/978-81-322-1958-3_4,

Genetic Algorithm (WSGA) and Nondominated Sorting Genetic Algorithm (NSGA-II) is investigated. Experimentation is carried out for Miller Operational Transconductance Amplifier (OTA) and Folded cascode OpAmp architecture.

Keywords Analog circuit design automation • CMOS operational amplifier • Device sizing • Multi-objective genetic algorithm • Miller OTA • Floded cascode operational amplifier

4.1 Problem Definition

The integrated circuit (IC) market growth has been explosive over the last decade. The design of complex ICs has been possible due to technology improvements and design automation tools. The design automation of analog integrated circuit is inevitable considering the recently emerging discipline such as System on Chip (SOC) design. The design of the analog portion of a mixed-signal integrated circuit often requires a large fraction of the overall design time, despite the fact that the analog circuit is often relatively a small portion of the overall circuit. So, it is necessary to develop Computer Aided Design (CAD) tools to automate the analog circuit synthesis process. There is a need for high-performance analog electronic circuit, which satisfies a large number of specifications. An achievement of fast, low-power, and low-area integrated circuits is the major requirement in the electronic industry.

It is not easy to predict the behavior of analog circuits when compared to digital circuits of similar complexity. Automation strategies used for digital synthesis cannot be applied for analog circuits; therefore, mixed-signal design has gained importance. Typical analog synthesis process consists of two main steps: topology selection and transistor sizing, before going down to the final layout (Fujii and Shibatu 2001; Hershenson et al. 1998; Maulik et al. 1992).

Analog synthesis process is classified under knowledge-based approach and optimization-based approach. The two distinct analog circuit synthesis approaches are shown in Fig. 4.1. Knowledge-based approaches use circuit equations, which require an expert understanding of the performance of the system. The formulation of circuit equations takes a long time, but once this is done, solutions can be found for solving the equations. Optimization-based systems generally use a search algorithm in conjunction with a circuit simulation engine to find a constraint-satisfying solution. This approach formulates the analog circuit design problem as an optimization problem and employs optimization technique to minimize or maximize a set of objectives while satisfying a set of constraints (Nye et al. 1988; Chang et al. 1992; Ochotta et al. 1996; Onodera et al. 1990; Gielen et al. 1990; Krasnicki et al. 1999). It consists of an optimization-driven search module and an evaluation module. At every iteration of the search, the performances of the circuit need to be evaluated for proper convergence to the desired specification.

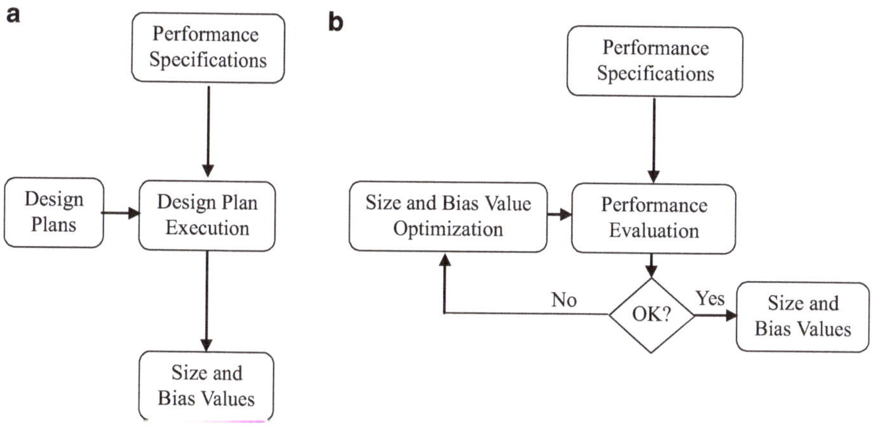

Fig. 4.1 Analog circuit synthesis (**a**) The knowledge-based approach (**b**) The optimization-based approach

Application of evolutionary computing technique in the domain of electronics has shown a potential that is attracting researchers and electronic circuit manufacturers. Memetic Single-Objective Evolutionary Algorithm (MSOEA) achieves higher global and local search ability by combining operators from different standard evolutionary algorithms (Liu et al. 2009a, b, 2014; Alpaydin et al. 2003). By integrating operators from the differential evolution algorithm, from the real-coded genetic algorithm, operators inspired by the simulated annealing algorithm, and a set of constraint handling techniques, MSOEA specializes in handling analog circuit design problems with numerous and tight design constraints. The method has been tested through the sizing of several analog circuits. The results show that design specifications are met and objective functions are highly optimized.

The Operational Amplifier (OpAmp) is the most versatile and widely used building blocks in analog electronics. The CMOS OpAmp design problem is a complex and tedious task, which requires many compromises to be made between conflicting objectives. The performance of an OpAmp is characterized by a number of parameters such as, gain, power, slew rate, and area. These performance parameters are determined by transistor dimensions, bias current, and compensation capacitance values. The design of CMOS OpAmp is translated to a multi-objective optimization task. Evolutionary algorithm is applied to the OpAmp design because of the multi-objective nature of the design problem and large search space. The evolutionary algorithm performs cell sizing, i.e., search for transistors sizes, biasing current, and compensating capacitance values to meet a set of specifications. Synthesis of CMOS OpAmp circuit is attempted using Weighted Sum Genetic Algorithm (WSGA) and Nondominated Sorting Genetic Algorithm (NSGA-II) (Deb et al. 1999).

4.2 Operational Amplifier Design

4.2.1 Miller OTA Architecture

The Miller Operational Transconductance Amplifier (OTA) circuit shown in Fig. 4.2 is a two-stage OpAmp. The first stage consists of a differential stage with PMOS transistor input devices M_1 and M_2 and the current mirror M_3 and M_4 acting as an active load. The second stage is a simple CMOS inverter with M_6 as driver and M_7 acting as active load. CMOS inverter output is connected to the output of the differential stage by means of a compensation capacitance C_C (Razavi 2001). Compensation capacitance actually acts as a miller capacitance. For most of the frequency ranges, the output impedance is low. It provides high gain and high output swing and is very suitable for low-voltage applications where few transistors can be stacked to provide sufficient gain.

4.2.1.1 Computation of Objectives

The Miller OTA with objectives area, power, slew rate, and gain bandwidth (GBW) is considered for the design. The objectives are determined using Eqs. 4.1, 4.2, 4.3, and 4.4 (Allen and Holberg 1987). The slew rate is determined using the bias current fixation as in Eq. 4.1; the OpAmp area has been estimated using Eq. 4.2:

$$I_5 \text{ or } I_{bias} = \text{Slew rate} \times (C_c) \tag{4.1}$$

$$A = \Sigma WL + 1,000\ C_C \tag{4.2}$$

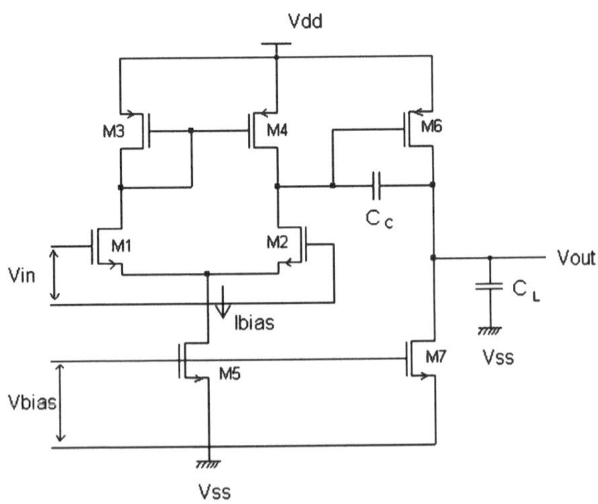

Fig. 4.2 Miller OTA schematic diagram

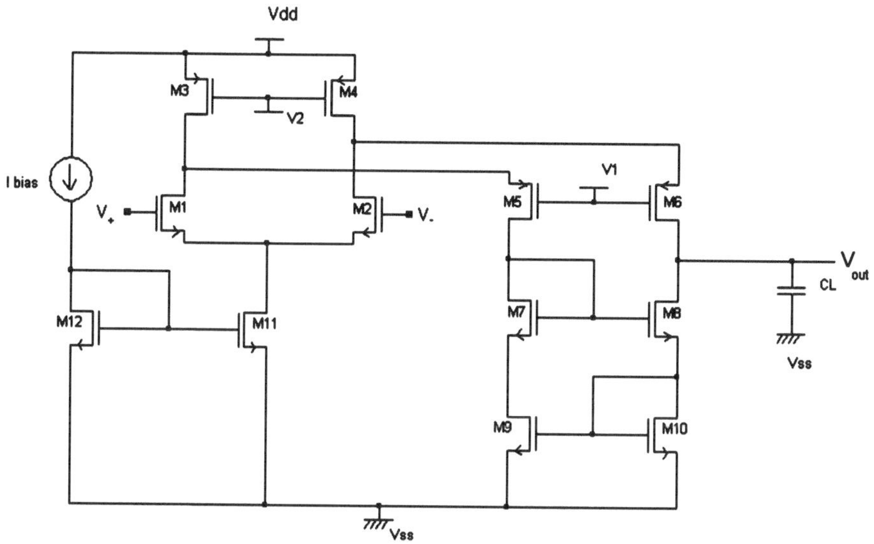

Fig. 4.3 Folded cascode OpAmp schematic diagram

where W and L represent the width and length of the transistors and C_c is the compensation capacitance in the second stage.

The power dissipation is computed using Eq. 4.3:

$$P = (V_{dd} - V_{ss})\,(I_5 + I_6) \tag{4.3}$$

where I_5 is the bias current and I_6 current through Transistor M6.

The gain bandwidth is computed using Eq. 4.4:

$$GBW = \frac{g_{m1}}{2\pi C_c} \tag{4.4}$$

where $g_{m1} = \sqrt{2K'_n I_5 \left(\frac{W}{L}\right)_1}$ and K'_n is the transconductance parameter

4.2.2 *Folded Cascode Amplifier Architecture*

Folded cascode amplifier is basically a single gain stage and its gain is high because the gain is determined by the product of input transconductance and output impedance. The schematic diagram of folded cascode OpAmp is shown in Fig. 4.3. It consists of a differential input and single-ended output stage and is a self-compensating OpAmp. The current mirror used in the circuit is cascode current mirror. The use of these mirrors results in high output impedance for the mirror and maximizing the dc gain of the OpAmp (Allen and Holberg 1987; Laker and Sansen 1994). The differential pair input transistors are n-channel transistor M_1 and M_2 and the cascode transistor consisting of M_5 and M_6 are p-channel transistors. The current mirror is composed of a pair of transistors M_7, M_8 and M_9, M_{10}. The bias circuit is constructed using transistors M_3, M_4, M_{11}, and M_{12}.

4.2.2.1 Computation of Objectives

The objectives of folded cascode amplifier is determined using the Eqs. 4.5, 4.6, 4.7, and 4.8 (Allen and Holberg 1987). The slew rate for the folded cascode amplifier is determined using bias current fixation as in Eq. 4.5:

$$I_{bias} = \text{slew rate} \times C_L \tag{4.5}$$

where C_L is the load capacitance.

The folded cascode OpAmp area has been estimated as the product of width and length of the transistors:

$$A = \Sigma WL \tag{4.6}$$

where W and L represent the width and length of the transistors.

The power dissipation is computed using Eq. 4.7:

$$P = (V_{dd} - V_{ss})\,(I_3 + I_{12} + I_{10} + I_{11}) \tag{4.7}$$

where I_3, I_{12}, I_{10}, and I_{11} are the currents in respective transistors M_3, M_{12}, M_{10}, and M_{11}.

The gain bandwidth is computed using Equation 4.8:

$$GBW = \frac{g_{m1}}{C_L} \tag{4.8}$$

where $g_{m1} = \sqrt{2K'_n I_{bias}\left(\frac{W}{L}\right)_1}$

4.3 Multi-objective Genetic Algorithm for Operational Amplifier Design

Weighted Sum Genetic Algorithm (WSGA) solves the multi-objective optimization (Michalewicz and Janikow 1991; Goh and Li 2001; Goldberg 1989) problems by combining the multiple objectives into a scalar cost function, ultimately making the problem single-objective before optimization. The Nondominated Sorting Genetic

W1	L1	W2	L2	W3	L3	W4	L4	W5	L5	I_{bias}	C_C

Fig. 4.4 Encoding structure of Miller OTA

W1	L1	W2	L2	W3	L3	W4	L4	W5	L5	W6	L6	I_{bias}

Fig. 4.5 Encoding structure of folded Cascode OpAmp

Algorithm (NSGA-II) method aims to determine the trade-off surface, which is a set of nondominated solution points, known as Pareto-optimal or noninferior solutions.

4.3.1 Circuit Representation for Miller OTA

The OpAmp design parameter is represented in a chromosome integer string so that each string element serves as pointer to the actual value of the OpAmp design parameter optimized and can be of any length depending on the number of parameters to be optimized. The encoding scheme for the Miller OTA is shown in Fig. 4.4.

The differential input transistor of Miller OTA is constrained to be equally sized. The transistor dimension, biasing current, and the compensation capacitance have been allowed to take hundred different values with reasonable minimum value, based on the technology. This encoding procedure, results in a search space of the order of 10^{30} possible solutions.

4.3.2 Circuit Representation for Folded Cascode OpAmp

The chromosome representation of the folded cascode architecture is shown in Fig. 4.5.

Folded cascode architecture is implemented with the transistor width and length limit set to 5–15 μm and I_{bias} set at 100 μA.

4.3.3 WSGA-Based OpAmp Design

The main challenge of applying GA or any other optimization technique to OpAmp design problem is the multi-objective nature of the same (Van Veldhuizen and Lamont 2000; Kruiskamp and Leenaerts 1995).The problem of multi-objective optimization concerns the need to integrate vectorial performance measures with

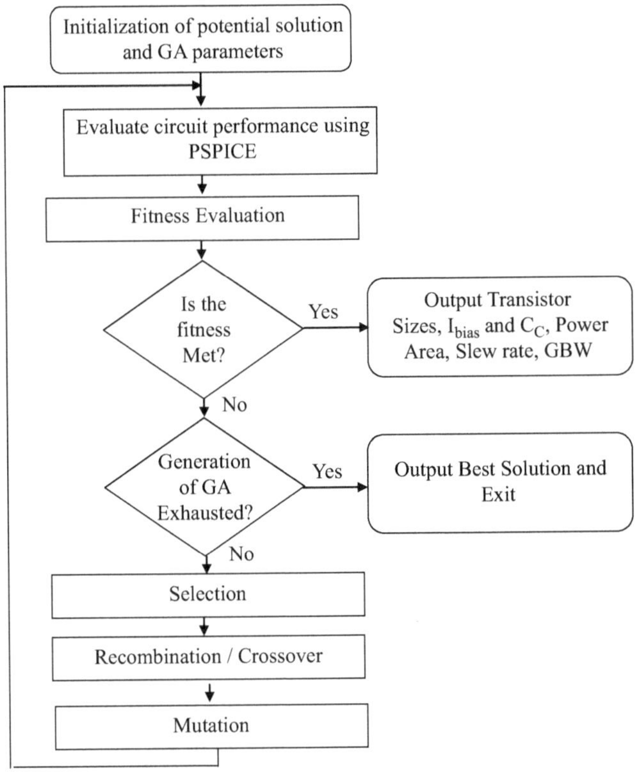

Fig. 4.6 OpAmp design using WSGA

the inherently scalar way in which most optimization techniques rewards individual performance. WSGA-based OpAmp design is shown in Fig. 4.6 (Shanthi and Bhuvaneswari 2009; Shanthi 2010).

4.3.3.1 Fitness Function

The fitness function for amplifier design is implemented as weighted sum of deviations between user specifications and value achieved by GA.

The proposed equation for the overall fitness is given by Eq. 4.9:

$$\text{Fitness} = \left(\frac{\text{K}}{(\text{eP} + \text{eA} + \text{eS} + \text{eGBW})}\right) \tag{4.9}$$

where, eP, eA, eS, and eGBW refer to the error in power, area, slew rate, and GBW as given by the Eqs. 4.10, 4.11, 4.12, and 4.13, respectively:

$$eP = \left(\frac{(P\text{-}P_{spec})}{P_{spec}}\right) \tag{4.10}$$

$$eA = \left(\frac{(A\text{-}A_{spec})}{A_{spec}}\right) \tag{4.11}$$

$$eS = \left(\frac{(S_{spec}\text{-}S)}{S_{spec}}\right) \tag{4.12}$$

$$eGBW = \left(\frac{(G_{spec}\text{-}G)}{G_{spec}}\right) \tag{4.13}$$

where P_{spec}, S_{spec}, A_{spec}, and G_{spec} are the specified values of the four parameters. P, S, A, and GBW are the obtained values through GA of the four parameters power, slew rate, area, and gain bandwidth. The actual error, i.e., the difference between the value obtained and the value specified is divided by the specified value, in order to normalize the errors in all the four objectives. Equal weights have been assigned to the four objectives. The fitness is inversely proportional to the total error and, hence, the equation for fitness is given by inverting the sum of the four errors magnified by the factor K. Here, K is taken as 100.

As the power and area are to be minimized, the corresponding error is taken as the difference between the obtained value and the specified value. On the other hand, slew rate and GBW are to be maximized and so, the error in this case is the difference between the specified value and the obtained value. It is ensured that the error is always positive.

4.3.4 Experimental Results of WSGA Method

The WSGA technique is applied to optimize the Miller OTA and folded cascode OpAmp architectures. The WSGA is used to manipulate the transistor sizes, biasing current I_{bias}, and the compensation capacitance C_C. The bounds used for the transistor width and length, I_{bias} and C_C, are given in Eqs. 4.14, 4.15, 4.16, and 4.17, respectively, for Miller OTA (Zebulum et al. 1998):

$$5\ \mu m < \text{width} < 105\ \mu m \text{ in steps of } 1\ \mu m \tag{4.14}$$

$$5\ \mu m < \text{length} < 105\ \mu m \text{ in steps of } 1\ \mu m \tag{4.15}$$

$$1.5\ \mu A < I_{bias} < 2.5\ \mu A \text{ in steps of } 0.01\ \mu A \tag{4.16}$$

$$0.1\ pF < C_C < 10\ pF \text{ in steps of } 0.1\ pF \tag{4.17}$$

A summary of the objectives and the desired values for Miller OTA is presented in Table 4.1. The supply voltages V_{DD} and V_{ss} are set to 5 V and 0 V, respectively.

Table 4.1 Design objectives for Miller OTA

No.	Objectives	Desired values
1	Power	≤200 μW
2	Slew rate	≥3.0 V/μs
3	Area	≤5,000 μm^2
4	GBW	≥3.5 MHz

Table 4.2 lists the parameter values of the ten best solutions generated by the WSGA for Miller OTA. The tabulated results are obtained by running the program 10 times, each time with different initial seed values. The population size is set as 10 and the code is run for 200 generations. In effect, 20,000 cells are sampled by the GA. The crossover probability is set to 0.87 and the mutation probability to 0.13. The method of crossover adopted is one-point crossover and the method of reproduction used is tournament selection with elitism.

Two-point and uniform crossover operators are implemented for Miller OTA OpAmp and the results obtained are given in Tables 4.3 and 4.4, respectively. Table 4.5 shows the comparison of the results. Higher fitness value is obtained for both roulette wheel and tournament selection with two-point crossover operator. High value of crossover probability is chosen to get new offspring. The mutation probability is set at 0.05 and the role of mutation in GA has been that of restoring lost or unexplored genetic material into the population to prevent the premature convergence of the GA to suboptimal solutions.

Table 4.6, shows that simple Miller OTA circuit synthesized by WSGA follows with the results obtained using equations in all aspects. OpAmp synthesized by WSGA features less than half of the power dissipation observed in the equation-based design, at the expense of large area. A graph showing the optimized values for the four objectives power, slew rate, area, and gain bandwidth against the number of generation for Miller OTA is shown in Fig. 4.7.

4.3.4.1 Experimental Results for Folded Cascode OpAmp

Folded cascode architecture is implemented with the transistor width and length limit set to 5–15 μm and I_{bias} set at 100 μA. The design objectives for folded cascode architecture are shown in Table 4.7. Comparison between crossover operators implemented for folded cascode amplifier is illustrated in Table 4.8. The crossover probability is set to 0.87 and the mutation probability to 0.13. Tournament selection operation is chosen. Higher fitness value is achieved with two-point crossover operator.

4.4 Operational Amplifier Design Using NSGA-II

The flowchart for NSGA-II-based OpAmp design is shown in Fig. 4.8. Initially, a random parent population P_t is created of size N. At first, the binary tournament selection, recombination, and mutation operators are used to create a child

Table 4.2 Ten best solutions generated by WSGA using one-point crossover

No.	W1 (μm)	L1 (μm)	W2 (μm)	L2 (μm)	W3 (μm)	L3 (μm)	W4 (μm)	L4 (μm)	W5 (μm)	L5 (μm)	I_{BIAS} (MA)	C_C (PF)	Power (MW)	SR (V/μs)	Area (μM^2)	GBW (MHz)	Fitness
1	12	24	47	37	50	8	5	31	31	41	1.83	4.7	211.8	2.980	4,696	3.577	1,505
2	12	24	47	37	13	46	5	52	32	27	1.54	4.7	206.3	2.937	4,895	3.577	1,801
3	52	6	47	37	50	8	5	36	32	41	1.70	4.7	210.5	2.990	4,835	3.577	1,692
4	12	24	46	37	30	8	5	52	32	41	1.74	4.7	210.7	2.986	4,947	3.577	1,627
5	12	33	47	37	31	8	5	52	32	36	1.68	4.7	209.6	2.975	4,699	3.577	1,684
6	12	24	47	37	44	8	5	74	32	31	1.51	4.7	209.6	3.009	5,071	3.577	1,533
7	17	24	47	37	23	8	5	52	32	41	1.51	4.7	209.6	3.009	4,755	3.577	1,526
8	12	24	47	37	54	8	5	52	33	41	1.51	4.7	209.4	3.005	5,052	3.577	1,655
9	12	24	47	37	49	9	5	52	33	40	1.64	4.7	209.9	2.989	5,037	3.577	1,572
10	12	24	47	37	67	10	5	9	33	40	1.51	4.7	209.2	3.002	5,065	3.577	1,606

Table 4.3 Ten best solutions generated by WSGA using two-point crossover

No.	W1 (μm)	L1 (μm)	W2 (μm)	L2 (μm)	W3 (μm)	L3 (μm)	W4 (μm)	L4 (μm)	W5 (μm)	L5 (μm)	I_{BIAS} (MA)	C_C (PF)	Power (MW)	SR (V/μs)	Area (μM^2)	GBW (MHz)	Fitness
1	12	30	47	37	28	9	5	42	33	40	1.51	4.7	209.2	3.002	4,847	3.489	2,279
2	12	24	47	37	49	9	5	52	33	40	1.64	4.7	209.9	2.989	5,037	3.489	1,572
3	18	24	47	37	23	8	5	52	21	63	1.51	4.7	212.1	3.060	4,814	3.489	1,818
4	12	24	47	37	54	8	5	52	33	41	1.51	4.7	209.4	3.005	5,052	3.489	1,655
5	17	24	47	37	28	8	5	52	32	41	1.51	4.7	220.9	3.009	4,835	3.489	2,210
6	12	24	47	37	50	8	5	36	32	41	1.54	4.7	209.7	3.006	4,787	3.489	2,181
7	12	24	47	37	50	8	5	31	35	41	1.54	4.7	209.2	2.995	4,860	3.489	1,973
8	12	24	47	37	50	8	5	104	12	41	1.54	4.7	212.4	3.060	4,647	3.489	1,782
9	21	24	47	37	50	8	5	22	31	41	1.54	4.7	209.9	3.009	5,038	3.489	1,743
10	12	24	47	37	35	10	5	42	33	40	1.66	4.7	210.0	2.987	4,755	3.489	2,033

Table 4.4 Ten best solutions generated by WSGA using uniform crossover

No.	W1 (μm)	L1 (μm)	W2 (μm)	L2 (μm)	W3 (μm)	L3 (μm)	W4 (μm)	L4 (μm)	W5 (μm)	L5 (μm)	I_{BIAS} (MA)	C_C (PF)	Power (MW)	SR (V/μs)	Area (μM^2)	GBW (MHz)	Fitness
1	14	21	60	51	11	8	18	21	9	47	1.54	8.5	200.6	2.824	5,003	3.541	1,600
2	14	15	50	5	44	8	18	21	9	34	1.54	9.7	200.6	2.824	4,736	3.541	1,865
3	14	15	60	57	11	8	18	21	9	47	1.54	8.5	200.9	2.830	4,835	3.541	1,878
4	20	30	44	34	6	19	63	11	18	48	2.40	6.0	219.6	3.032	5,174	3.541	1,763
5	20	19	44	34	6	19	63	11	32	18	2.31	9.7	204.1	2.740	4,446	3.541	1,686
6	20	44	34	6	19	63	17	17	18	18	1.77	8.3	213.3	3.032	5,312	3.541	1,618
7	21	18	44	34	6	83	28	11	36	18	1.58	9.7	193.8	2.677	4,512	3.541	1,578
8	20	19	44	31	6	83	28	11	36	18	1.92	9.7	215.4	3.043	4,384	3.541	1,675
9	20	30	44	34	6	82	63	11	18	18	1.77	6.0	213.1	3.027	5,450	3.541	1,637
10	20	30	44	34	6	40	63	11	18	18	2.11	6.0	214.7	2.992	5,006	3.541	1,780

Table 4.5 Comparison between crossover operators

Objectives	Roulette wheel selection			Tournament selection		
	Single-point crossover	Two-point crossover	Uniform crossover	Single-point crossover	Two-point crossover	Uniform crossover
Power (μW)	210.0	260.4	234.0	199.3	209.2	200.6
Slew rate (V/μS)	3.019	2.947	2.873	2.797	3.002	2.824
Area (μm^2)	4,707	4,880	5,088	4,964	4,847	4,660
GBW (MHz)	3.644	3.575	3.424	3.577	3.489	3.541
Maximum fitness value	1,542	2,264	1,730	1,688	2,279	1,748

Table 4.6 Comparisons of results between the equation-based and WSGA-based methods for Miller OTA

Objectives	Equation-based result	WSGA-based result
Slew rate	3.78 V/μs	3.002 V/μs
I_B	2.50 μA	1.5 μA
C_P	1 pF	4.7 pF
Power	527.8 μw	209 μw
Area	1,929 μm^2	4,847 μm^2
GBW	2 MHz	3.489 MHz
C_L	10 pF	10 pF

population Q_t of size N. A combined population $R_t = P_t \cup Q_t$ is formed. The individuals in the population are sorted based on nondomination into each front.

Each individual in the front is assigned rank values. Individuals in first front are given a rank value of 1 and individuals in second are assigned rank value as 2 and so on. The first front being completely nondominant set (Coello 1999; Deb et al. 1999; Fonseca and Fleming 1998) in the current population and the second front being dominated by the individuals in the first front only and the front goes and so on. As the entire population size of R_t is 2 N, not all the fronts can be accommodated in the N slots available in the new parent population. The fronts that cannot be accommodated are deleted. When the last allowed front is considered, there may be more solutions in the last front than the remaining slots to be filled in the new population. Instead of randomly discarding some solutions from the last front, a niching strategy is used to include the solutions of the last front that are present in the least-crowded region in that front. A new parameter called crowding distance is calculated for each individual. The crowding distance is a measure of how close an individual is to its neighbors. Large average crowding distance will result in better diversity in the population. An individual is selected if the rank is lesser than the other or if crowding distance is greater than the other. The new population P_{t+1} of size N are used for selection, crossover, and mutation to create new offspring population Q_{t+1} of size N. The process is repeated until the desired fitness is reached and the W/L value, I_{bias} and compensation capacitance value satisfying the objectives are displayed.

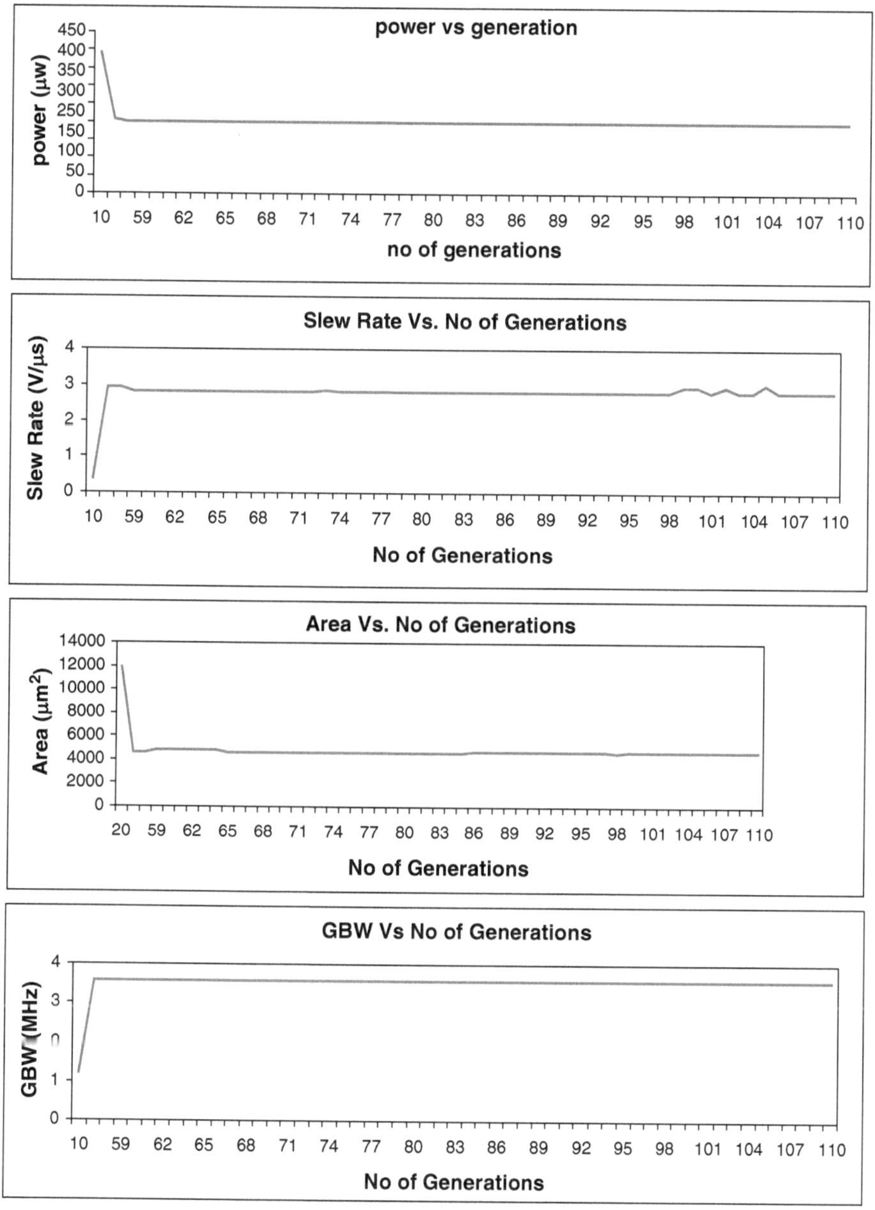

Fig. 4.7 Four objectives considered by evolutionary process for the optimization of Miller OTA

Table 4.7 Design objective for folded cascode OpAmp

No.	Objectives	Desired values
1	Power	≤800 μW
2	Slew rate	≥15.0 V/μs
3	Area	≤1,000 μm
4	GBW	≥70 MHz

Table 4.8 Comparison between crossover operators

	Tournament selection		
Objective	Single-point crossover	Two-point crossover	Uniform crossover
Power (μW)	761.2	761.7	752.7
Slew rate (V/μs)	14.897	14.95	14.88
Area (μm^2)	980	904	978
GBW (MHz)	66	67	65
Maximum Fitness value	2,551	2,838	2,108

4.4.1 Multi-objective Fitness Function

Fitness function estimation is challenging as the OpAmp specification are conflicting in nature. OpAmp must exhibit low power dissipation and high gain bandwidth at the same time. The fitness function is implemented as follows:

$$\text{Power} = \text{Power}_{\text{min desired}} - \text{Power}_{\text{actual}} \tag{4.18}$$

$$\text{Area} = \text{Area}_{\text{min desired}} - \text{Area}_{\text{actual}} \tag{4.19}$$

$$\text{Slew rate} = \text{Slew rate}_{\text{actual}} - \text{Slew rate}_{\text{min desired}} \tag{4.20}$$

$$\text{GBW} = \text{GBW}_{\text{actual}} - \text{GBW}_{\text{min desired}} \tag{4.21}$$

This ensures that all the objectives at least meet the lower bound of the desired specification.

The fitness function for power (Eq. 4.18) resolves to negative value if the power has exceeded the minimum allowable amount. Similarly, the fitness function for GBW (Eq. 4.21) resolves to negative value, if the minimum desired GBW has not been obtained. The algorithm seeks to maximize the deviation between the desired and the actual value, which in turn maximizes or minimizes the objective depending on its formulation. The fitness function implemented ensures that all the objectives satisfy the minimum desired value.

4.4.2 Simulation Results Obtained Using NSGA-II Algorithm

Simulation is carried out in a Pentium IV processor clocking at 1.4 GHz in Windows environment. In order to fully automate the design and minimize the use of human expertise, PSPICE is used to evaluate the circuit parameters.

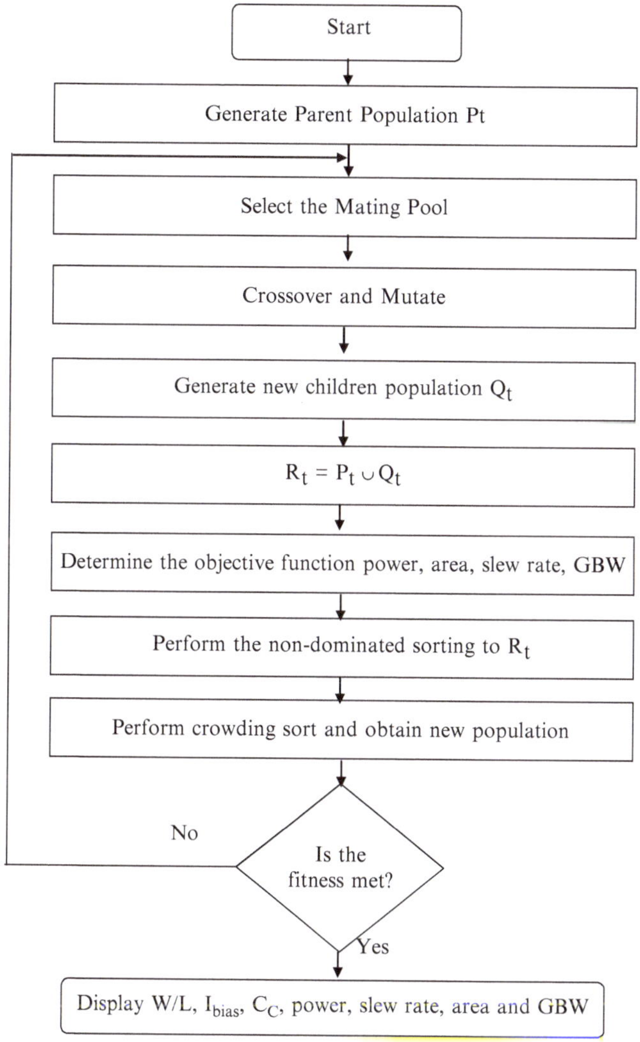

Fig. 4.8 NSGA-II based OpAmp design

Population size of 10 is used with a crossover probability and mutation probability of 0.95 and 0.05, respectively.

The effectiveness of using the NSGA-II multi-objective algorithm is demonstrated for Miller OTA and folded cascode amplifier optimizing conflictive objectives. The Pareto-fronts for the two conflicting objectives, power and slew rate for Miller OTA are shown in Fig. 4.9a, b. The Pareto-fronts are obtained from generation 5 and generation 40 of the evolution process. In the beginning of the evolution, the objectives are scattered. Many of the individuals do not satisfy

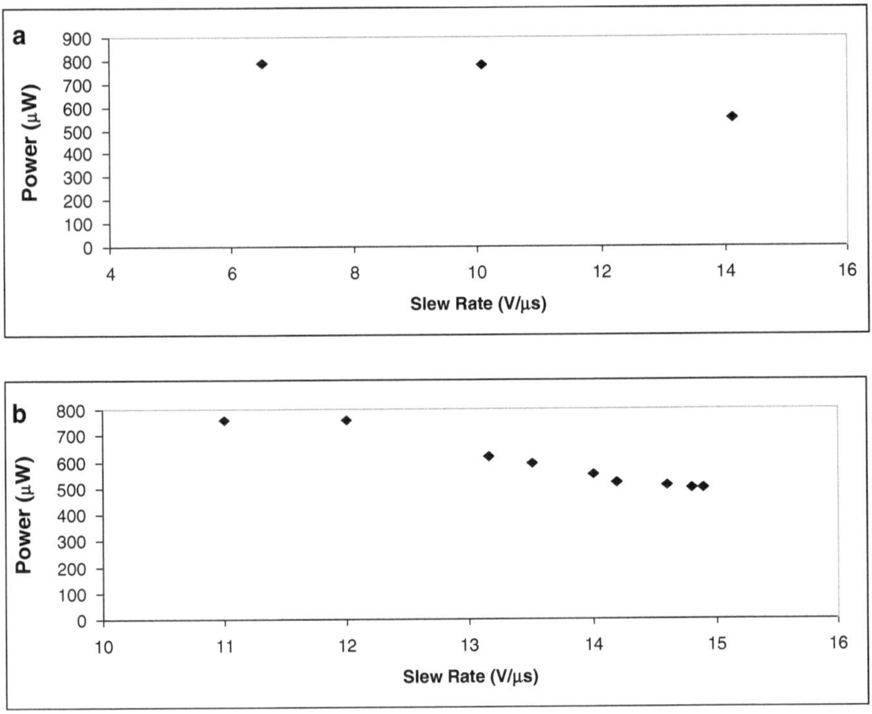

Fig. 4.9 Pareto-fronts for a two objective optimization (Conflicting) (**a**) Generation 5 (**b**) Generation 40

both the objectives. As the evolution progresses, that is, in generation 40, the Pareto-front is more evenly distributed, showing that more individuals satisfy both the objectives. The Pareto-fronts obtained for Miller OTA, considering power and area for optimization is illustrated in Fig. 4.10.

Table 4.9 presents the performance comparison of NSGA-II-based results with equation-based results for one of the individuals obtained in 40th generation for Miller OTA and folded cascode OpAmp. It can be seen that the performance of the circuits using NSGA-II provides lower power dissipation than the equation-based values.

The device sizes for folded cascode OpAmp is shown in Table 4.10 for the solutions obtained in generation 40. The results obtained for folded cascode amplifier using NSGA-II is shown in Table 4.11 for 20 generations and the results obtained using WSGA is shown in Table 4.12 for 200 generations. The simulation results obtained using NSGA-II shows the convergence results of the algorithm takes a run time of 19 min in 20 generations itself.

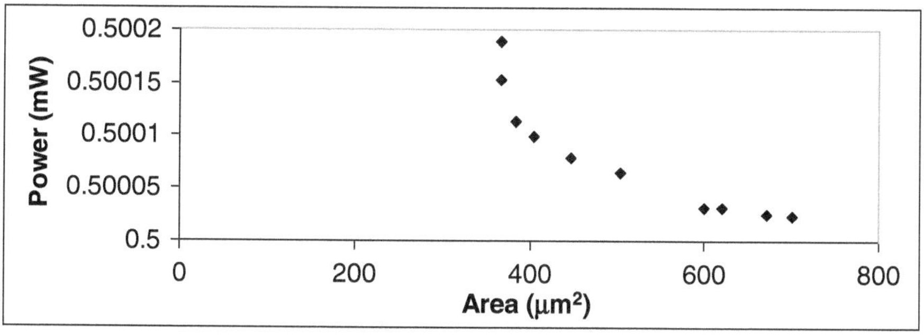

Fig. 4.10 Pareto-fronts for two objectives optimization (Minimization)

Table 4.9 Comparisons of results between the equation-based and NSGA-II-based methods

		Objectives			
Architecture	Methodology	Power (μw)	Area (μm^2)	Slew rate (V/μs)	GBW (MHz)
Miller OTA	Equation based	527	1,929	3.78	2
	NSGA-II	157	4,326	6.5	3.1
Folded cascode	Equation based	800	780	10	70
	NSGA-II	500	973	15	74

Table 4.10 Device sizes of the folded cascode OpAmp solutions obtained using NSGA-II

Variable	Value
W1	8 μm
L1	9.9 μm
W2	9.9 μm
L2	9.9 μm
W3	9.9 μm
L3	8.54 μm
W4	9.5 μm
L4	9.7 μm
W5	9.7 μm
L5	8.89 μm
W6	5 μm
L6	9.8 μm
I_{bias}	149 μA

Table 4.11 Optimal solutions using NSGA-II (20 generations)

Output	Power (MW)	Area (MM^2)	Slew rate (V/MS)	GBW (MHZ)
Solution 1	740.5	897	14.8	73.2
Solution 2	750.0	893	14.9	73.2
Solution 3	749.0	934	14.9	71.8
Solution 4	700.0	837	14.0	71.9
Solution 5	750.0	609	14.98	74.9
Solution 6	700.0	959	14.009	73.8
Solution 7	709.6	765	14.18	73.6

Table 4.12 Optimal solutions using WSGA (200 generations)

Output	Power (MW)	Area (MM^2)	Slew rate (V/MS)	GBW (MHZ)
Solution A	763.5	962	14.95	67
Solution B	762.8	990	14.95	68.8
Solution C	751.3	954	14.95	66
Solution D	748.3	950	14.95	65.7
Solution E	761.9	984	14.95	67.9
Solution F	761.8	1,014	14.95	66.3
Solution G	748.6	910	14.95	69.1

4.5 Summary

A method to synthesize CMOS OpAmp based on WSGA and NSGA II algorithm is presented. Automation tool based on concepts of GA is developed using C language, which performs automatic synthesis of CMOS OpAmp. PSPICE is interfaced with C program in order to obtain parameter value needed to calculate fitness. Given a design plan, the developed application performs an efficient search over the design space. Miller OTA and folded cascode amplifier implementation provided solutions satisfying all the objectives close to the specifications. Elitist Nondominated Sorting Genetic Algorithm (NSGA-II) for synthesizing Miller OTA amplifier and folded cascode design provides multiple-optimal solutions for the transistor sizing, biasing current, and compensation capacitor within 20 generations.

References

Allen PE, Holberg DR (1987) CMOS analog circuit design. Rinchart and Winston Editors, Holt

Alpaydin G, Balkir S, Dundar G (2003) An evolutionary approach to automatic synthesis of high-performance analog integrated circuits. IEEE Trans Evol Comput 7(3):240–252

Chang H, Charbon E, Choudhur U, Demir A, Felt E, Liu E, Malavasi E, Sangiovanni-Vincentelli-A, Vassiliou I (1992) A top-down constraint-driven design methodology for analog integrated circuits. In: IEEE custom integrated circuit conference, Boston, pp 8.4.1–8.4.6

Coello CAC (1999) A comprehensive survey of evolutionary multi-objective optimization techniques. Knowl Inf Syst Int J IEEE Serv Cent 1(3):269–308

Deb K, Pratap A, Agarwal S, Meyarivan T (1999) A fast and elitist multi-objective genetic algorithm. NSGA-II, KanGAL Indian Institute of Technology, Kanpur, India

Fonseca CM, Fleming PJ (1998) Multi-objective optimization and multiple constraint handling with evolutionary algorithms – Part I: a unified formulation. IEEE Trans Syst Man Cybern Part A 28(4):26–37

Fujii N, Shibatu H (2001) Analog circuit synthesis by superimposing of sub-circuit. In: Proceedings of the international symposium on circuits and systems, vol 5(1), pp 427–430

Gielen GE, Sansen WMC, Walscharts HCC (1990) Analog circuit design optimization based on symbolic simulation and simulated annealing. IEEE J Solid-State Circuits 25:707–713

Goh C, Li Y (2001) GA automated design and synthesis of analog circuits with practical constraints. In: Proceedings of the 2001 congress on evolutionary computation, vol 1(1), pp 170–177

Goldberg DE (1989) Genetic algorithms in search, optimization and machine learning. Addison-Wesley, New York

Hershenson M, Boyd SP, Lee TH (1998) GPCAD: a tool for CMOS OpAmp synthesis. IEEE Trans Comput Aided Des :296–303

Krasnicki M, Phelps R, Rutenbar RA, Carley LR (1999) Maelstrom: efficient simulation-based synthesis for custom analog cells. In: Proceedings of design automation conference, New York, USA, pp 945–950

Kruiskamp W, Leenaerts D (1995) DARWIN: CMOS OpAmp synthesis by means of genetic algorithm. In: 32nd annual design automation conference, New York, USA, pp 433–438

Laker KR, Sansen W (1994) Design of analog integrated circuits and systems. McGraw Hill, New York

Liu B, Fernández FV, Gielen G, Castro-López R, Roca E (2009a) A memetic approach to the automatic design of high-performance analog integrated circuits. ACM Trans Des Autom Electron Syst 14(3), Article 42

Liu B, Wang Y, Yu Z, Liu L, Li M, Wang Z, Lu J, Fernández F (2009b) Analog circuit optimization system based on hybrid evolutionary algorithms. Integr VLSI J 42(2):137–148

Liu B, Fernández FV, Gielen G (2014) Automated design of analog and high frequency circuits. In: A computational intelligence approach. Springer, Berlin, Heidelberg, ISBN: 978-3-642-39161-3

Maulik PC, Flynn M, Allstot DJ, Carley LR (1992) Rapid redesign of analog standard cells using constrained optimization techniques. In: IEEE custom integrated circuit conference pp 8.1.1–8.1.3

Michalewicz Z, Janikow CZ (1991) Handling constrains in genetic algorithms. In: Proceedings 4th international conference on genetic algorithms. Morgan Kaufmann Publishers, San Diego, CA, San Mateo, pp 151–157

Nye W, Sangiovanni-Vincentelli A, Tits AL (1988) DELIGHT SPICE: an optimized-based system for the design of integrated circuits. IEEE Trans Comput Aided Des 7(4):501–519

Ochotta ES, Rutenbar RA, Carley LR (1996) ASTRX/OBLX: synthesis of high-performance analog circuits. IEEE Trans Comput Aided Des 15:273–293

Onodera H, Kanbara H, Tamaru K (1990) Operational amplifier compilation with performance optimization. IEEE J Solid-State Circuits 25:466–473

Razavi B (2001) Design of analog CMOS integrated circuit. The McGraw-Hill Companies Inc., Boston. ISBN 0-07-118815-0

Shanthi M (2010) Certain investigations on synthesis of CMOS operational amplifier using evolutionary techniques" M.S (By research) thesis, Anna university, Chennai

Shanthi M, Bhuvaneswari MC (2009) Design of CMOS operational amplifiers using multiobjective genetic algorithm. Int Eng Technol J Adv Comput 3(1):008–011

Van Veldhuizen DA, Lamont GB (2000) Multi-objective evolutionary algorithms: analyzing the state-of-the-art. IEEE Trans Evol Comput 8(2):125–147

Zebulum RS, Pacheco MA, Vellasco M (1998) Synthesis of CMOS operational amplifiers through genetic algorithm. In: Int conference on microelectronics and packaging, Brazil, pp 1–4

Chapter 5
Design Space Exploration for Scheduling and Allocation in High Level Synthesis of Datapaths

M.C. Bhuvaneswari, D.S. Harish Ram, and R. Neelaveni

Abstract Increasing design complexity and shrinking lead times have led to automation of the VLSI design flow at higher levels of abstraction. High-level synthesis (HLS) is that phase in system design which deals with the translation of behavioral descriptions in high-level languages such as C or graphical representations in the form of data flow graphs (DFGs) and control data flow graphs (CDFGs) into an equivalent register transfer level (RTL) netlist. The typical objectives to be optimized during HLS such as delay, area, and power are mutually conflicting thereby necessitating the rapid exploration of the entire design space to identify solutions with different trade-offs among the objectives. This is called design space exploration (DSE). The NP-complete nature of HLS problems requires heuristic approaches which are capable of yielding near optimal solutions. Population-based meta-heuristics such as genetic algorithms (GAs) and particle swarm optimization (PSO) are ideal candidates for DSE since they are capable of generating a population of trade-off solutions in a single run. The application of multi-objective GA and PSO approaches for optimization of power, area, and delay during datapath scheduling and allocation phases of HLS is discussed in this chapter. A novel metric-based approach for estimating the potential of a schedule to yield low-power bindings is also proposed.

Keywords High level synthesis • Behavioral synthesis • Multi-objective evolutionary algorithms • Low power design • Datapath scheduling and allocation

---

M.C. Bhuvaneswari • R. Neelaveni
Electrical and Electronics Engineering, PSG College of Technology, Coimbatore, India
e-mail: mcb@eee.psgtech.ac.in; rnv@eee.psgtech.ac.in

D.S. Harish Ram (✉)
Electronics and Communication Engineering, Amrita Vishwa Vidyapeetham University, Coimbatore, India
e-mail: ds_harishram@cb.amrita.edu

M.C. Bhuvaneswari (ed.), *Application of Evolutionary Algorithms for Multi-objective Optimization in VLSI and Embedded Systems*, DOI 10.1007/978-81-322-1958-3_5,

5.1 Introduction

The ever increasing complexity of chip designs and shrinking lead times have spawned the development of methodologies for automating the VLSI design flow at higher levels of abstraction. Electronic System Level (ESL) design deals with the different methodologies and tools employed for exploring and evaluating different design alternatives at the system and algorithmic levels. ESL also enables designers to identify optimal hardware-software partitions for meeting lower level design objectives such as area, delay, power, and throughput. Incorporating these objectives in the early stages of the design flow facilitates better optimization of the designs due to the limited flexibility available to the designer, as the design gets more and more committed to silicon in the later stages of the VLSI flow. At higher levels of abstractions, the designs are specified in terms of transaction level models. Computational tasks are specified using algorithmic descriptions in high level languages or graphically in the form of Control Data Flow Graphs (CDFGs).

In datapath intensive algorithms, the graphical depiction is in the form of a Data Flow Graph (DFG). High Level Synthesis (HLS) is the phase in ESL which deals with the automatic translation of an algorithmic description to a Register Transfer Level (RTL) netlist. The HLS process involves synthesis of both the datapath and controller hardware. However, most commercial tools are unable to handle both datapath and control intensive algorithms with equal efficiency (Martin and Smith 2009).

This Chapter discusses evolutionary approaches based on Weighted Sum Genetic Algorithm (WSGA), Weighted Sum Particle Swarm Optimization (WSPSO), and Non-dominated Sorting Genetic Algorithm–II (NSGA- II) for concurrent power, area, and delay optimization in Datapath Synthesis. Section 5.2 gives an overview of datapath synthesis. Section 5.3 reviews the related work reported in the literature. The implementation details of the proposed techniques are described in Sect. 5.4. The results of evaluating the techniques on standard DFG benchmarks are analyzed in Sect. 5.5. Section 5.6 concludes the chapter.

5.2 Datapath Synthesis

This Chapter deals with the High Level Synthesis of datapaths from DFGs. An example DFG that depicts a datapath operation is shown in Fig. 5.1 (Harish Ram et al. 2012).

The nodes of the DFG represent the datapath operations that are part of the algorithm depicted by the DFG. Thus, node *a* is an addition node whereas node *d* represents a multiplication operation. The edges in the DFG correspond to data transfers between the nodes. Thus the results of nodes *a* and *b* are required for the

Fig. 5.1 Data flow graph

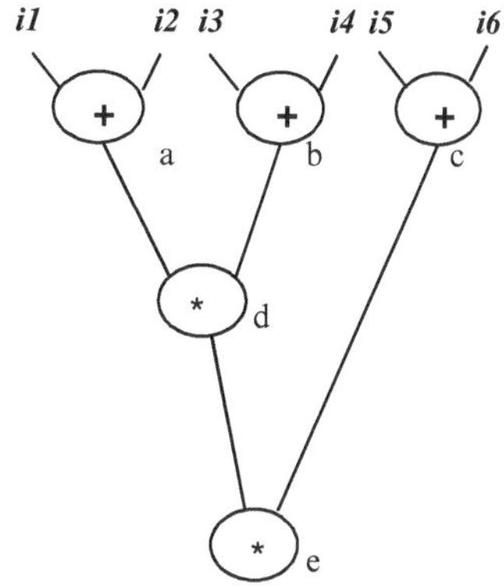

execution of node *d*. Hence if the clock period of the system is restricted to the delay of a Functional Unit (FU), then the execution of the DFG entails multiple clock cycles. Thus registers are assigned for storing the node results which are consumed by the successor nodes at their scheduled time step

A naïve approach for implementing the DFG would be to have a one to one correspondence between the DFG and the FUs allocated. Thus, for the DFG in Fig. 5.1 a total of three adders and two multipliers will be allocated. However, this approach understandably leads to a large area overhead especially as the number of nodes in the DFG scales up. Moreover, the large number of FUs and associated glue logic will lead to unacceptably high dynamic as well as leakage power numbers. The mutual exclusivity between nodes executing identical operation is exploited for sharing of FUs between nodes.

Datapath Synthesis comprises of three sub-phases viz., *scheduling*, *allocation* and *binding*. *Scheduling* determines the time step in which a particular node in the DFG is to be executed. *Allocation* assigns Functional Units (FUs) for execution of the operations represented by the various DFG nodes. The registers required for the intermediate storage of the results of execution of each node are also allocated in this phase. *Binding* assigns an FU for executing each node. The specific storage resource for each node is also assigned in this phase. It should be noted that there is no specific order in which these sub-phases are carried out. However, the order does influence the RTL that is eventually synthesized.

A possible schedule for the DFG in Fig. 5.1 is shown in Fig. 5.2a. Scheduling algorithm routinely exploit the *mobility* among the nodes to optimize FU allocations. For the example DFG, the node *c* can be moved to the second time step without affecting the schedule length of 3 time steps. This reduces the number of adders to 2 from 3 since only two addition nodes *a* and *b* execute concurrently in time step 1.

The post-binding RTL for the schedule in Fig. 5.2a is shown in Fig. 5.2b. It consists of the two adders and a multiplier and multiplexers (MUX) for routing

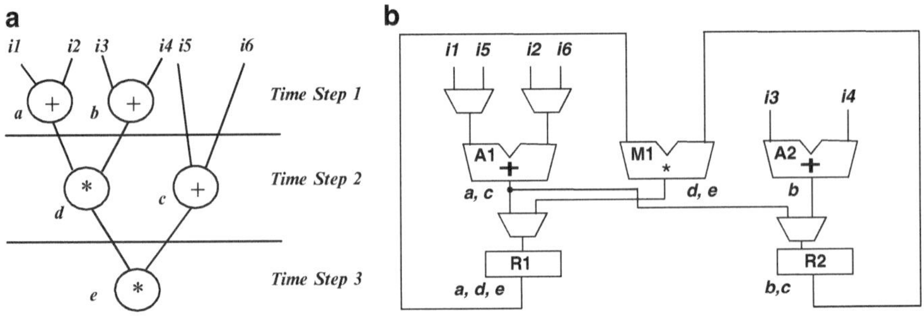

Fig. 5.2 (a) Schedule for DFG in Fig. 5.1. (b) RTL for the Schedule in Fig. 5.2a

the appropriate inputs to the shared FUs. The MUX input lines driving the FUs are selected by the control logic which is typically a state machine or sequencer. Registers are assigned for the intermediate storage of results of the nodes. In Fig. 5.2a, the adder *A1* is assigned to the nodes *a* and *c*. The node *b* which executes concurrently with *a* is bound to adder *A2*. The register *R1* is assigned respectively to *a*, *d*, and *e*. The nodes *b* and *c* are bound to register *R2*. A pair of MUXes are required at the input of *A1* and its inputs are driven by the input pair (*i1*, *i2*) when node *a* is executed and the pair (*i5*, *i6*) when node *c* is executed. However, MUXes are not required for multiplier M1 though nodes *d* and *e* share this FU. The predecessor nodes of *d* are *a* and *b* which are read from registers *R1* and *R2* respectively. The inputs to node *e* are the results of nodes *d* and *c* which are assigned registers *R1* and *R2* as before eliminating the need for MUXes to route the inputs. The register inputs are also driven by MUXes since the nodes sharing these resources execute in different FUs. Thus, it is evident that the binding of nodes into FUs and registers influences not only the FU resource cost but the cost of steering logic such as multiplexers as well. The use of multiplexers has an adverse effect on area and switching power in FPGA targets since MUXes are not efficiently implemented in FPGAs (Casseau and Le Gal 2009). Thus, interconnect binding also plays a significant role in HLS targeted at FPGAs (Casseau and Le Gal 2009; Chen et al. 2010).

5.3 Related Work

It has been proven that scheduling and binding of DFGs under constraints such as FU allocation is NP-complete, thus necessitating heuristic approaches to arrive at near optimal solutions (Gerez 2000). Early work in behavioral synthesis focused on *constructive approaches* which schedule one node at a time. Nodes are taken up for scheduling based on some criteria such as mobility. In one such approach called *force directed scheduling* (Paulin and Knight 1989), nodes with non-zero mobility are assigned time steps that minimize the *force* of a particular schedule. The force is a metric that indicates the utilization of FUs in different time steps. A lower value for the force metric indicates a balanced utilization of FUs in the different cycles. This is a greedy approach vulnerable to local minima.

An Integer Linear Programming (ILP) approach is combined with retiming for power optimization in (Chabini and Wolf 2005). A game-theoretic approach for power optimization of a scheduled DFG is described in (Murugavel and Ranganathan 2003). The functional units are modeled as bidders for the operations in the DFG with power consumption as the cost. The algorithm does not scale well for larger number of functional units since the complexity increases exponentially as the number of players (bidders).

The presence of multiple conflicting objectives to be optimized makes the use of evolutionary techniques such as Genetic Algorithm (GA) and Particle Swarm Optimization (PSO) attractive for the HLS problem. These meta-heuristics being population based, are capable of identifying a large pool of solutions that involve different trade-offs among the objectives in a single run.

A Weighted Sum GA (WSGA) approach for area and delay optimization in datapath synthesis is reported in (Krishnan and Katkoori 2006). The authors use WSGA for simultaneous scheduling and functional unit allocation for behavioral synthesis of DFGs with area and delay minimization as the optimization criteria. Sengupta and Sedaghat (2011) use a weighted GA approach for design space exploration during DFG scheduling similar to (Krishnan and Katkoori 2006) but with a more refined cost function. The authors incorporate power in the cost function in addition to area and delay cost. A more accurate model for delay based on actual cycle time in a pipelined execution environment is used instead of simply relying on cycle times of operations.

Particle Swarm Optimization (PSO) proposed by Eberhart and Kennedy is another evolutionary approach widely used to solve optimization problems. The optimal scheduling problem in HLS is somewhat analogous to the traveling salesman problem (TSP). In both cases, an evolutionary approach necessitates a chromosome encoding which specifies the *order* in which scheduling of nodes (cities to be visited in case of TSP) is carried out. These problems differ in their cost functions. TSP seeks to minimize the distance travelled whereas in HLS, the objective is to optimize area, delay or power or all the three depending on the case. In addition, the precedence of execution of each node is to be considered during HLS. Wang et al. (2003), proposed a PSO methodology for TSP. The authors describe a technique for generating new scheduling strings using the swap operator. An encoding scheme for efficient convergence of the particles during PSO has been proposed by (Lin et al. 2008). Jie et al. (2010) described a scheme for reducing the scale of the problem for large number of nodes without affecting exploration of the search space. In (Abdel-Kader 2008), a PSO engine is proposed for resource-constrained instruction scheduling for minimization of total schedule time. Hashemi and Nowrouzian (2012) proposed a PSO based approach for simultaneous scheduling and binding of filter datapaths with area and delay minimization as the objectives. The technique uses a fixed allocation for the functional units. A multi-chromosome is used for respectively specifying the scheduled time step and module binding for each node in the DFG. The binding field has an entry corresponding to the order in which inputs are assigned to the FU. This provides scope for generating solutions with optimized multiplexer counts. A weighted sum cost function incorporating the area of the binding and the delay of the implementation is used to evaluate the fitness of individual solutions.

The weighted sum approach essentially converts a multi-objective problem into a single objective one. For many problems, these methods are inefficient in exploring the entire solutions space and converging towards the true Pareto Front (Deb 2008). Several *true* multi-objective approaches have been proposed (Deb 2008) which employ the concept of non-dominance. The main challenges in applying true multi-objective techniques is to identify an efficient means of generating nondominated solutions from a population besides reducing the effort of computing the fitness of individual solutions. In (Pilato et al. 2010), which is an extended version of (Ferrandi et al. 2007), a design space exploration scheme using NSGA-II for area and time optimization during HLS is proposed. The authors use a scheduling based chromosome encoding similar to the one described in this chapter and also an alternative binding based encoding to stress binding optimization. Linear regression models are used for computation of area and delay cost. The notion of *fitness inheritance* is used to limit the effort of computing the fitness of individuals. Here, the cost of an individual solution is inherited from another individual based on the distance between the two. Thus, actual cost needs to be computed only for a fraction of the individuals in a population. Power consumption is not addressed in this work.

In Schafer and Wakabayashi (2011), a method called Divide and Conquer Exploration algorithm (DC ExpA) is proposed for generating Pareto optimal solutions during behavioral synthesis. A high-level description is parsed and divided into *clusters* which exhibit interdependent operations. Design Space Exploration is performed on each cluster in an independent fashion to obtain Pareto optimal sub-sets. These are merged to generate the final solutions after accounting for the sharing of variables between different clusters. A set of non-dominated solutions is obtained from the merged individuals using a greedy approach. The authors report improvement in run times with comparable solution quality compared to a simulated annealing engine. Also, improvements in solution quality are reported with a reduction in run time in comparison to a similar divide and conquer approach. However, the method appears to be vulnerable to local minima since the scope for optimization among different clusters is not explored. This may have to be further studied vis a vis population based approaches such as GA.

The above mentioned drawback is addressed by the authors in Schafer and Wakabayashi (2012). Here, a high level description is grouped into sets of explorable constructs to which random attributes are applied to create a candidate solution. Initially a set of such solutions is synthesized and the area and delay cost of each solution is evaluated using actual synthesis and latency estimation using a cycle accurate simulator. A predictive model is derived for these solutions using a machine learning approach and the model is compared with the actual latency and the area numbers of the solutions generated so far. If the error is less than a threshold, the model is used for fitness evaluation of a subsequent GA run which seeks to guide the solutions towards the Pareto front. If the predictive model exhibits an error higher than the threshold, a random set of solution is generated once again.

This exercise is repeated till the prediction error reduces below the threshold value. The methodology is reported to show improvement in runtimes but a trade-off is reported between the maximum error threshold and solution quality. Lower thresholds lead to better solution quality albeit at the expense of longer run times necessitated by the larger iterations required to refine the predictive model. The methodology does not take into account power consumption during the design space exploration phase.

Similar work is reported in (Zuluaga et al. 2012), where a Gaussian Regression machine learning approach is used to develop a predictive model for evaluating candidate solutions during HLS with the dual objective of minimizing area and maximizing throughput. The model is enhanced with a technique for smart selection of samples with convergence towards the Pareto optimality as the main objective. The authors demonstrate the superiority of the method over a Matlab based regression approach and a Gaussian approach with random selection. Xydis et al. (2013) proposed a design space exploration scheme for HLS based on Response Surface Method (RSM). RSM is a multi-objective exploration tool that is capable of predicting the metrics of candidate solutions based on an analytical model that is derived by means of a *training phase*. The training phase is carried out by synthesizing solutions at the RTL level and computing delay and area metrics to create the analytical model of the solution surface. The Response surface exists as a meta-layer to guide and refine the synthesizer for converging towards non-dominated solutions without expensive synthesis and characterization steps. The authors claim improvements in solution quality compared to NSGA-II and Multi-objective Simulated Annealing in terms of closeness to the true Pareto Front. Also, speedup of the exploration phase is reported compared to the aforementioned methods. However, the method does not incorporate power as a metric during the exploration phase.

The application of NSGA-II for datapath scheduling with concurrent power, area and delay optimization is addressed in this work. The techniques reported in the literature primarily address area and delay. Power optimization is rarely addressed. The NSGA-II based technique provides an effective means for rapid design space exploration of datapath schedules with different trade-offs in power, area, and delay costs.

5.4 Multi-objective Evolutionary Approaches to Datapath Scheduling and Allocation[1]

Evolutionary approaches have been investigated for the datapath synthesis problem in works of (Krishnan and Katkoori 2006) and (Ferrandi et al. 2007). In (Krishnan and Katkoori 2006), the authors have proposed a weighted sum GA for area and delay optimization in datapath HLS. A true multi-objective approach based on

[1] This section is reproduced from the paper "A Novel Framework for Applying Multi-objective GA and PSO-based Approaches for Simultaneous Area, Delay And Power Optimization In High-Level Synthesis Of Datapaths", VLSI Design Journal, 2012 by D. S. Harish Ram et al. under the creative commons attribution license from M/s Hindawi Publishing Corporation.

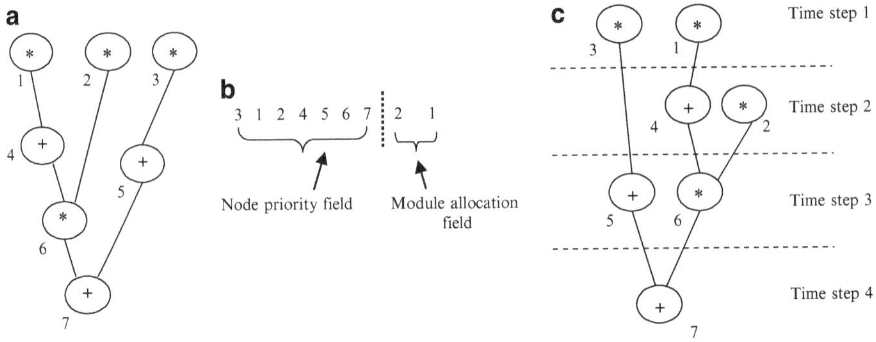

Fig. 5.3 (**a**) Unscheduled DFG (**b**) Chromosome (**c**) Scheduled DFG

NSGA-II is reported for area and delay optimization in datapath HLS in (Ferrandi et al. 2007) with a more accurate cost function which incorporates binding information. Power is rarely addressed in the multi-objective techniques reported in the literature. Moreover, the cost function evaluation is made expensive due to the necessity of low-level simulations and characterizations. Also, the cost function models used are library dependent. A multi-objective evolutionary framework is described in this chapter for concurrent power, area, and delay optimization in datapath synthesis. A metric based approach for predicting the likelihood of a schedule to yield low power bindings is used as the power cost. Actual power numbers are not required. The framework is applied to Weighted Sum GA (WSGA), NSGA-II, and Weighted Sum PSO (WSPSO) approaches for the scheduling and allocation problem.

5.4.1 Encoding Scheme

A multi-chromosome encoding scheme reported in (Krishnan and Katkoori 2006) is used in the proposed framework. The same encoding is used for both GA and PSO. In this scheme a chromosome has a *node scheduling priority field* and a *module allocation field* as shown in Fig. 5.3b. The structure of the chromosome is such that simultaneous scheduling of a DFG and functional unit allocation can be carried out. The DFG nodes are scheduled using a list scheduling heuristic (Krishnan and Katkoori 2006; Gerez 2000). The nodes are taken up for scheduling in the order in which they appear in the chromosome. The module constraint is described in the *module allocation field*. The respective allocation constraints for the multipliers and adders are specified in this field. If additional FUs such as subtractors are present, the module allocation field will have more terms.

As an example, an unscheduled DFG and a corresponding chromosome encoding are shown respectively in Fig. 5.3a, b. The scheduling starts with the first time step. Node 3 which appears first in the chromosome is scheduled in the first time step itself. Also, Node 1 is scheduled to be executed in time step 1 since it appears next in the chromosome. Node 2 which is next cannot be scheduled in the first time step because only two multipliers are available as specified in the module allocation field. Therefore, Node 2 has to wait till the next time step. Node 4 which performs addition cannot be scheduled in this time step since its predecessor node i.e., Node 1 has just been scheduled. The result of the execution of this node will be available only in the next time step. Similarly, nodes 5, 6 and 7 cannot be scheduled since the results of their predecessor nodes are not ready. So, a move to the second time step is done since no further scheduling is possible in the current time step. In this step, nodes 2 and 4 which are not scheduled in the previous time step are scheduled for execution. This sequence continues till all nodes are scheduled. It has to be noted that in a valid string, the precedence relationship between the nodes has to be maintained. For instance, in the chromosome shown in Fig. 5.3b, the nodes 1 and 2 which are the predecessors for node 6 appear before node 6 in the priority list. The scheduled DFG is shown in Fig. 5.3c.

5.4.2 Cost Function Computation

The fitness evaluation of each chromosome is carried out as described in (Krishnan and Katkoori 2006) for area and delay. The potential of a given schedule to yield low power bindings is determined from the compatibility graphs for the FUs and registers extracted from the schedule using a method described in (Kursun et al. 2005). Actual power numbers are not computed.

5.4.2.1 Area and Delay Cost

The delay or length L is the number of time-steps or clock cycles needed to execute the schedule. For example $L=4$ for the schedule in Fig. 5.3c. The area cost A of the schedule is based on the total FUs and registers required to bind all the nodes in the DFG. The former is known from the chromosome itself (e.g., 2 multipliers and 1 adder for the DFG in Fig. 5.3c). The latter is obtained by determining the total number of registers required for the DFG implementation using the left edge algorithm (Gerez 2000). The total area is expressed as the total number of transistors required to synthesize the FUs and registers to a generic library.

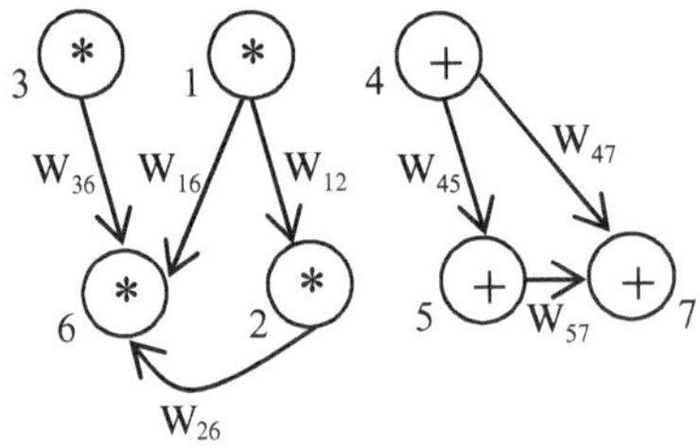

Fig. 5.4 Compatibility graph

5.4.2.2 Power Cost

The potential of an individual schedule to yield a low-power binding solution is found out using a set of metrics described in (Kursun et al. 2005). Here a *compatibility graph (CG)* is extracted from the scheduled DFG corresponding to the schedule obtained from the node priority field in each chromosome of the population. The compatibility graph will have an edge between two nodes if they are *compatible*. Two nodes are said to be compatible if they can be bound to the same FU or register. For instance, nodes 1 and 2 in the DFG in Fig. 5.3c are FU compatible since they execute the same operation (multiplication) and their lifetimes do not overlap. On the other hand, nodes 3 and 1 though executing the same operation are not compatible since they execute concurrently. The compatibility graph for the schedule given in Fig. 5.3c, is given in Fig. 5.4. Separate CGs are created for the registers and FUs. The edge weights between two nodes represent the *switching cost* when these two nodes are executed one after the other in the same FU (Chang and Pedram 1995).

The power-related metrics (Kursun et al. 2005) are based on the edge weights of the compatibility graph and are defined as follows:

m_1 = Total number of edges in the graph,
m_2 = Average of edge weights for the lowest k % in the value range for each node where k is user-defined,
m_3 = Average of all edge weights (i.e., m_2 for $k = 100$ %). A DFG may yield a low-power binding if m_1 is high and m_2 and m_3 are low. The power cost function designed based on the compatibility graph metrics is given by Eq. (5.1)

$$P = \frac{n}{m_1} + m_2 + m_3 \tag{5.1}$$

where n is a tuning parameter and is chosen depending on the value of m_1 and has a value greater than m_1. The rationale for the choice of the power metrics is as follows. A higher edge count (m_1) in the CG indicates higher number of possible bindings and hence more likelihood of the schedule yielding low-power bindings. Thus, the power metric is made *inversely proportional* to m_1. Smaller values of CG edge weights indicate lesser switching cost and hence, lower power dissipation

Fig. 5.5 Crossover

Parent 1 ->	3	1	2	4	5	6	7	2	1	Crossover point
Parent 2 ->	1	2	4	6	3	5	7	2	2	
Offspring 1 ->	3	1	2	4	6	5	7	2	2	
Offspring 2 ->	1	2	4	6	3	5	7	2	1	

when a pair of compatible nodes execute on an FU. Thus, a schedule having a CG with smaller average edge weight has better potential to yield more power-aware bindings. This dependency is encoded in the second and third terms of the cost function, m_2 and m_3. A set of edge typical edge weights and the sample calculation of the power metric for the CG shown in Fig. 5.4 are shown below:

$w_{12} - 0.40$; $w_{16} = 0.40$; $w_{36} = 0.51$; $w_{26} = 0.45$; $w_{45} = 0.52$; $w_{47} = 0.38$; $w_{57} = 0.53$;
$m_1 =$ **7** (the number of edges)
$m_2 =$ **0.4280** (average of lower k % of edge weights. The value of k is chosen as 70 %. Thus the lowest five edge weights are averaged)
$m_3 =$ **0.4557** (average weight of all the edges)
Power metric $\text{P} = \frac{8}{7} + 0.4280 + 0.4557 =$ **2.0266** (n is chosen as 8)

5.4.3 *Crossover*

The single-point crossover technique reported in (Krishnan and Katkoori 2006) is used. In this method, the precedence relationships between the nodes in the node priority field of the multi-chromosome are maintained. The technique is shown in Fig. 5.5.

The parent string is retained intact till the crossover point. For instance, in the example shown in Fig. 5.5, crossover is carried out from the fifth position in the string. Thus, in offspring 1, the sub-string 3 1 2 4 is replicated from Parent 1. From location 5 onwards, the Parent 2 string is traversed and those nodes in the list that are *not present* are filled in the same order in which they appear in Parent 2. For instance, the nodes 1, 2 and 4 are already present in the sub-string 3 1 2 4 taken from Parent 1. But the node 6 is not present and hence is filled in the location 5. The node 3 which appears next is already present whereas 5 which appears next to 3, is not present and is filled in. The procedure is continued till the entire string is generated. The same sequence is repeated with the Parent 2 for generating the Offspring 2.

For the module allocation field, the module allocations which appear after the crossover point are swapped. In Fig. 5.5, the module allocation strings are 2 1 and 2 2 respectively for parent *Parent 1* and *Parent 2*. After crossover, the module allocation strings become 2 2 and 2 1 respectively.

5.4.4 Mutation

Mutation also is based on the technique described in (Krishnan and Katkoori 2006). For the node priority field, a node in the list is chosen at random and is moved without affecting the node precedence constraints. As an example, consider the same chromosome described previously, i.e. the string 3 1 2 4 6 5 7 | 2 1. Let the node 5 in location 6 be chosen for the mutation. Its predecessor is node 3 in the first slot and successor is the node 7 in the seventh slot. Thus the node can be moved to any slot from slot 2 to slot 6 without violating the precedence constraints. The actual slot to which the node is to be moved is chosen by a generating a random number between 2 and 6. If the random number generated is assumed as 3, then the node 5 is moved to slot 3 and the new chromosome after mutation becomes 3 1 **5** 2 4 6 7| 2 1 (mutated node is shown in bold).

For the module allocation field, the mutation is effected by simply incrementing or decrementing the number of functional units. For example the number of FU may be decremented in the above multi-chromosome. Thus, the final chromosome string becomes 3 1 **5** 2 4 6 7| **1 1** after mutation (mutated locations are shown in bold).

5.4.5 Weighted Sum GA for Multi-objective Datapath Synthesis

The weighted sum approach (Deb 2008) converts a multi-objective problem into a single objective GA. Each term in the cost function of this GA represents the cost of one objective to be optimized and is multiplied by a weight factor between 0 and 1. This weight is chosen depending upon the extent to which a particular objective is to be optimized. This approach is computationally simple. The fitness evaluation of each individual is based on a cost function value calculated using a weighted sum approach. The cost function described in (Krishnan and Katkoori 2006) is modified to include a power dissipation term described in the previous section and is given by Eq. (5.2).

$$C = w_1 L_s / L_{\max} + w_2 A / A_{\max} + w_3 P \tag{5.2}$$

where w_1, w_2 and w_3 are the weights of the area, delay and power terms respectively. The sum of the weights is always 1. The terms L_s and A are the length and area respectively of the implementation of the given schedule and P is the power metric which is computed using Eq. 5.1. The terms $A_{\max}$ and $L_{\max}$ are the maximum area and delay respectively among solutions in the current population. The power metric is not normalized since its range is bounded due to the fractional values of the edge weights and the reciprocal of the number of edges. A population size of 100 is used in the GA implementation. Elitism is used for preserving the best solutions in a generation. Binary tournament selection is used to identify parents for crossover. A crossover probability of 0.9 is used and the mutation probability is 0.2. The WSGA run is repeated with 10 different initial seed populations.

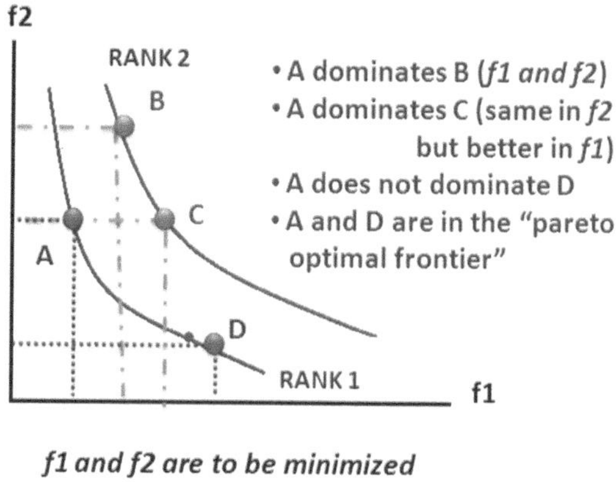

Fig. 5.6 Multi-objective solution space and Pareto Front

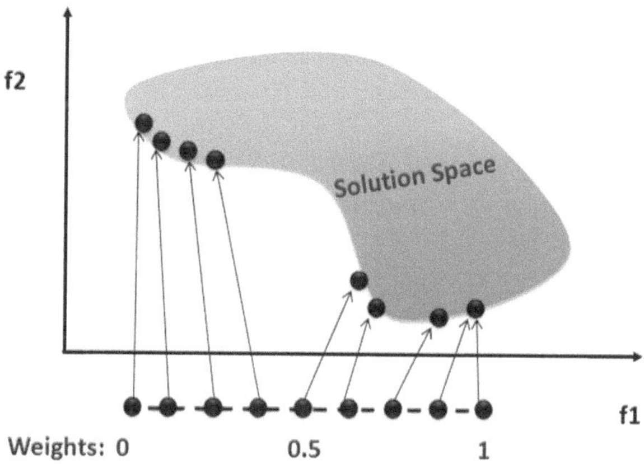

Fig. 5.7 Solutions for WSGA (Deb 2008)

5.4.6 Non-dominated Sorting Genetic Algorithm–II (NSGA-II) for Datapath Synthesis

In a *true multi-objective* algorithm, the solutions converge to the *true* Pareto front of *non-dominated* solutions. A solution is said to be *non-dominated* if no other solution can be found in the population that is better for *all the objectives*. A multi-objective solution space for a minimization problem involving two objectives and the Pareto front is depicted in Fig. 5.6. The weighted sum approach described in the previous section suffers from lack of diversity in that uniformly spaced weights do not result in a uniform spread of solutions as shown in Fig. 5.7. Figure 5.7 depicts a solution space for a minimization problem with two objectives *f1* and *f2*. The X-axis represents different values of *f1* for uniformly spaced values for the weight for *f2* (say, w_2). For $w_2 = 0$, *f1* has the most optimal value and *f2* the worst. For higher values of w_2, the objective *f2* shows progressively better values by trading off *f1*. But, it can be observed that the values of *f2* obtained by trading off *f1* are not uniformly spaced though the weight for *f2* (w_2) is increased in equal steps. Besides, proper assignment of weights and hence solution quality depends on knowledge of the solution space (Deb 2008).

Deb et al. proposed the "Non-Dominated Sorting GA-II" or NSGA-II (Deb et al. 2002), which is a true multi-objective GA. It uses the notion of crowding distance to ensure diversity among the solutions in a population. Initially, a random seed is created. Chromosome encoding and objective functions are same as WSGA. The cost of each objective for all the solutions in determined and they are classified into *Ranks* based on non-dominance. The Rank I individuals are fully non-dominated whereas those in Rank 2 are dominated by the Rank I individuals and so on. Each solution is assigned a fitness based on its Rank and *Crowding Distance*. The Crowding Distance is a measure of the uniqueness of a solution. Crossover and mutation are performed on the individuals using the method described in (Krishnan and Katkoori 2006). The parents and offspring in a particular generation are merged and the individuals for the next generation are selected based on the *crowding distance metric* (Deb et al. 2002; Deb 2008). Selection of individuals with higher crowding distance is favored for better diversity among solutions. A population size of 100 is used in each generation and the algorithm is run for 200 generations. A flow diagram depicting the NSGA-II methodology for DFG scheduling is shown in Fig. 5.8.

5.4.7 Weighted Sum Particle Swarm Optimization (WSPSO) for Multi-objective Datapath HLS

Particle Swarm Optimization (PSO) is an evolutionary approach proposed by Eberhart and Kennedy, inspired by the behavior of bird flocks or schools of fish. PSO like GA operates on a population of candidate solutions referred to as a *swarm*. Each solution in a swarm is called a *particle*. The swarm is made to evolve by applying a *velocity* to each particle. The best solution in each swarm is called the "particle best" or *pbest* and the best solution among all the swarms so far is the "global best" or *gbest*. When the swarm moves, each particle is subjected to a velocity which tends to propel it in the direction of *pbest* as well *gbest* with each direction being assigned a weight as modeled by Eq. (5.3)

$$v^i_{j,t+1} = wv^i_{j,t} + \eta_1 R_1 \left(p^i_{j,t} - x^i_{j,t}\right) + \eta_2 R_2 \left(g^i_{j,t} - x^i_{j,t}\right) \tag{5.3}$$

where $v^i_{j,\,t+1}$ = velocity of the ith particle for the t+1th iteration

$v^i_{j,t}$ = velocity of the ith particle for the tth iteration

The weight w assigns some *inertia* to the particle. The parameters η_1 and η_2 assign weights to the *pbest* and *gbest* variables i.e. the extent to which these positions influence the movement of the particle. The random values R_1 and R_2 introduce a degree of perturbation in the particle for better exploration of the search space.

5.4.7.1 Weighted Sum PSO (WSPSO) for DFG Scheduling

Scheduling of DFGs during datapath synthesis is a good candidate for application of PSO. This problem is somewhat analogous to TSP (Wang et al. 2003; Fang

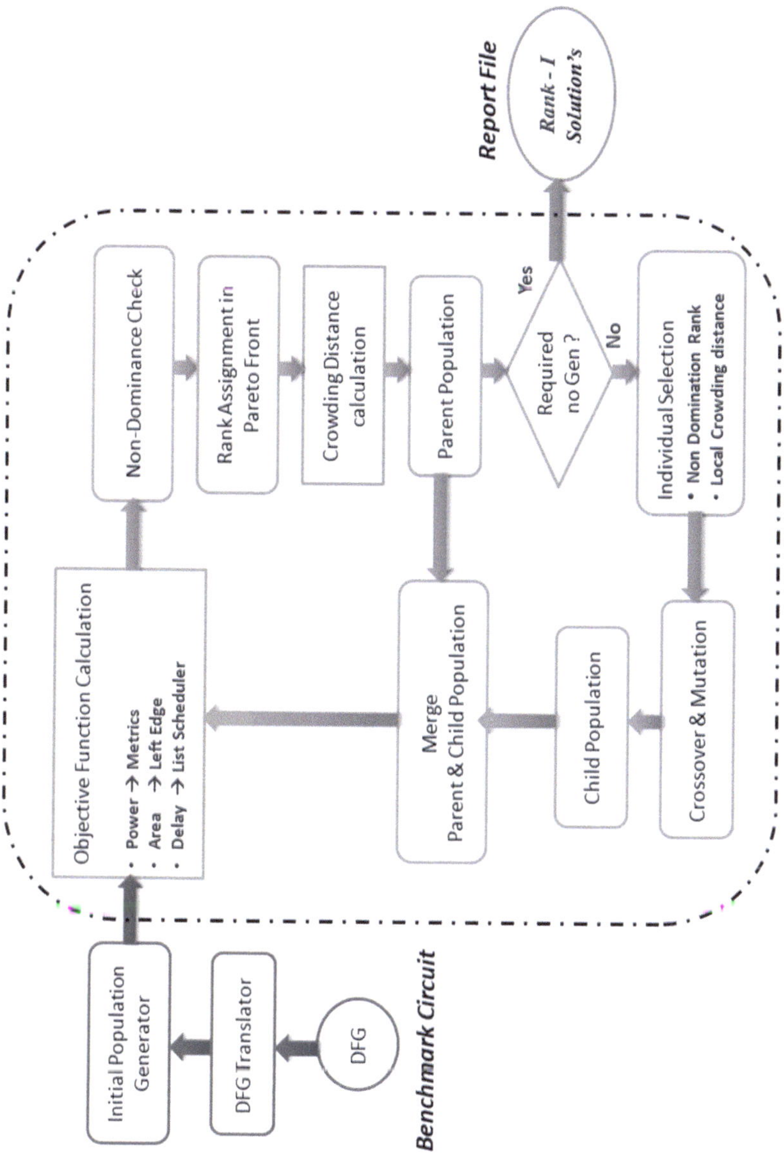

Fig. 5.8 NSGA-II based methodology for DFG scheduling

Fig. 5.9 Velocity in WSPSO

particle - 3 ¦ 1 2 4 5 6 7 | 2 1

gbest - 1 ¦ 2 4 6 3 5 7 | 2 2

new particle – 3 1 2 4 6 3 5 7 | 2 2

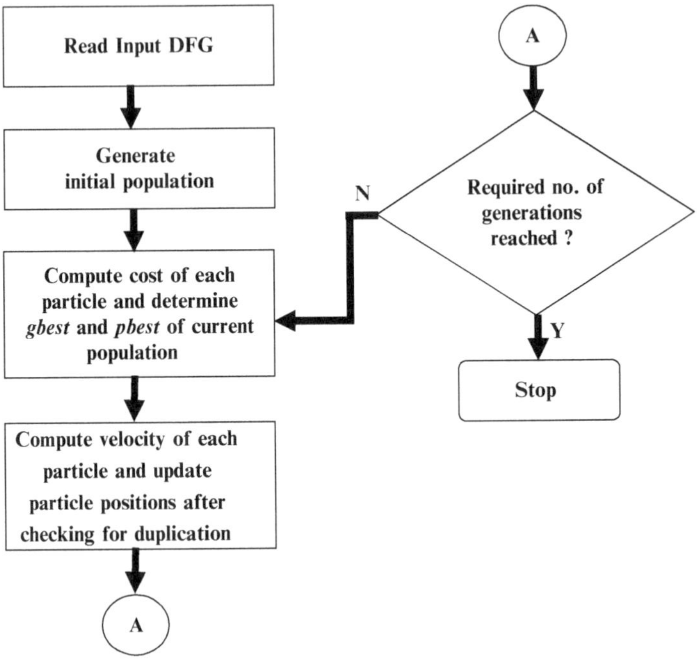

Fig. 5.10 PSO methodology for DFG scheduling

et al. 2007), albeit with precedence and module constraints. The multi-chromosome encoding introduced in Sect. 5.4.1 can be used for PSO also. In discrete problems, the notion of velocity is implemented by adapting the crossover technique (Krishnan and Katkoori 2006; Fang et al. 2007) for the PSO problem. A particle in the swarm is represented by a single multi-chromosome string representing a given DFG schedule and allocation. This particle is crossed over with the current particle that represents the *gbest* (or *pbest*) to implement the velocity function. The particle is crossed over with both the *gbest* and *pbest* as illustrated in Fig. 5.9.

The same approach is followed for *pbest* also. It can be seen that the degree of shift of the particle towards the *pbest* or *gbest* can be controlled by appropriately choosing the crossover location. If the location is in the beginning of the string then the shift is more.

Area, Delay, and Power costs of the schedule represented by the chromosome is computed using the method described in Sect. 5.4. The weighted sum approach outlined in Section Eq. 5.2 is used to determine the fitness of a particle represented by the multi-chromosome. A swarm size of 100 individuals is used. The total number of generations (swarms) is 70. The flowchart in Fig. 5.10 summarizes the WSPSO-based methodology used. The proposed approach is limited to DFGs and is not equipped for handling Control Data Flow Graphs (CDFGs) which are used for loop based algorithms. A PSO based heuristic for CDFG synthesis is presented in (Sengupta and Mishra 2014).

5.5 Results and Discussion[2]

The algorithms are coded in C and executed on an Intel i5-2400 CPU with a clock frequency of 3.10 GHz and 4 GB RAM.

5.5.1 Comparison of Power Aware WSGA with WSGA Without Power in the Cost Function (Krishnan and Katkoori 2006)

One of the contributions of this work is the introduction of a power metric that indicates at an early stage the *likelihood* of a schedule to yield a low power binding solution during the binding phase. This metric is added to the weighted sum cost function for area and delay optimization proposed in (Krishnan and Katkoori 2006). The efficacy of the modified GA with power is verified by evaluating the modified WSGA on various benchmarks and comparing with the results of WSGA method reported in (Krishnan and Katkoori 2006) which does not incorporate power in the cost function. The weight assigned to the power cost is 0.7 and the other objectives are assigned equal weights of 0.15 each.

It is observed that the modified GA methodology yielded schedules that had improved values of the power metric indicating higher likelihood of obtaining low power bindings. The results of the comparison are listed below in Table 5.1. An average reduction of around 9 % is observed in the power metric for the power aware GA run. The reduction is not significant for the IIR benchmark which has only 9 nodes. Hence, the search space of feasible schedules is limited. The HAL benchmark though having only 10 nodes shows a higher improvement of 12 % in the power numbers since it has higher mobility for the nodes and hence there is more number of feasible schedules. The power aware WSGA run yields better power optimal solutions for the FIR and MPEG benchmarks which have 23 and 28 nodes respectively, obviously due to the higher number of nodes and also more degrees of freedom in moving the nodes for generating different schedules.

[2] The Sects. 5.5 and 5.6 are reproduced from the paper "A Novel Framework for Applying Multi-objective GA and PSO-based Approaches for Simultaneous Area, Delay And Power Optimization In High-Level Synthesis Of Datapaths", VLSI Design Journal, 2012 by D. S. Harish Ram et al. under the creative commons attribution license from M/s Hindawi Publishing Corporation.

Table 5.1 Comparison of WSGA and power aware WSGA

	WSGA without power optimization (Krishnan and Katkoori 2006)			Power aware WSGA			
Benchmark	Power metric	Area (register units)	Delay (time steps)	Power metric	Area (register units)	Delay (time steps)	Reduction in power cost metric (%)
IIR	0.9062	42.9	5.44	0.8996	48.48	5.15	0.73
HAL	0.9911	38.33	4.74	0.8719	37.81	4.98	12
FIR	0.6487	42.73	11.63	0.6029	54.48	11.32	7
MPEG	0.6317	81.5	8.50	0.5246	107.5	9.00	17

Table 5.2 Metrics evaluating closeness to the Pareto-optimal front for IIR benchmark for WSGA and NSGA-II

Performance metric	WSGA	NSGA-II
Error ratio	0.7143	0
Generational distance	0.02708	0
Maximum Pareto optimal front error (MFE)	0.1833	0
Spacing	0.1046	0
Spread	1.0026	0
Weighted metric	0.5148	0

Table 5.3 Metrics evaluating closeness to the pareto-optimal front for HAL benchmark for WSGA and NSGA-II

Performance metric	WSGA	NSGA-II
Error ratio	0.7826	0
Generational distance	0.0050	0
Maximum Pareto optimal front error	0.1832	0
Spacing	0.0516	0
Spread	1.8600	0
Weighted metric	0.9360	0

5.5.2 Comparison of NSGA-II and WSGA

The quality of solutions obtained using a multi-objective algorithm must be assessed in terms of the spread of the solutions and the closeness of the individuals to the true Pareto front. The metrics described in (Deb 2008) for diversity, spread and Pareto front errors are computed for the Rank-I individuals obtained for the IIR filter and HAL benchmarks using WSGA and NSGA-II. For computing the quality metrics, the true Pareto front is obtained for these two benchmarks by exhaustive search. The metric values obtained for the WSGA and NSGA-II runs are listed in Tables 5.2 and 5.3. The *Error Ratio* is the fraction of solutions in the population of Rank I individuals which are not members of the true Pareto-optimal set, P^*. The *Generational Distance* metric gives a measure of the average distance of the solutions from P^*. The *Maximum Pareto Front Error* gives the distance of the solution which is farthest from the Pareto front. The values of the metrics, *Spacing*

Table 5.4 Comparison of WSGA and NSGA-II on standard benchmarks

	WSGA			NSGA-II		
Benchmark	Power metric	Area (register units)	Delay (time steps)	Power metric	Area (register units)	Delay (time steps)
IIR	0.9019	48.48	5.148	0.9063	35.41	5.88
HAL	0.9901	37.81	4.981	0.9923	39.91	4.75
DWT	0.7126	50.92	10	0.6821	48.95	10
FIR	0.6416	54.48	11.32	0.6312	43.58	11.23
MPEG	0.6714	107.5	9	0.6561	63.13	10.71
DCT	0.6086	79.06	12.18	0.6087	77.44	12.12

Table 5.5 Execution times in seconds

Benchmark circuit	WSGA	NSGA	% reduction
IIR	6	4	33.33
HAL	11	11	0
DWT	26	5	80.77
FIR	112	9	91.96
MPEG	48	23	52.08

and *Spread* indicate the diversity of solutions obtained. The *Weighted Metric* is the weighted sum of the closeness and diversity measures. These metrics should have small values for a good multi-objective algorithm. The results of comparing NSGA-II and WSGA in terms of the quality metrics are given below in Tables 5.2 and 5.3 respectively for the IIR and HAL benchmarks.

The above metrics indicate the superior quality of the NSGA-II solutions over those obtained by the Weighted Sum GA approach. All metrics have zero values for NSGA-II indicating that the solutions are *fully coincident* with the True Pareto front individuals.

The NSGA-II methodology is evaluated on several standard DFG benchmarks and the results are summarized below in Tables 5.4 and 5.5. Table 5.4 shows the average values of the delay, area, and power metrics obtained for the WSGA and NSGA II runs. Power dissipation is quantified in terms of the potential of the schedule to yield low power bindings as described in Sect. 5.4. Area is computed as the total gate count of the FUs and registers when synthesized to a generic library and is normalized to the register gate count as register units. Delay is expressed in terms of the total time steps required for completing each schedule. The average values for the NSGA-II run are either better than WSGA or comparable in most of the cases. For the smaller DFGs (IIR and HAL), the NSGA-II results do not exhibit an appreciable improvement in the *average* values. This is because of the fewer number of feasible solutions. Since NSGA-II searches the solution space more efficiently, the effect of trade-offs between the three objectives is more pronounced. Another aberration that is noticeable is in the case of the MPEG Motion Vector benchmark where a substantial reduction in average area is noticed at the expense of higher average delay. This DFG is rich in multiplication nodes whose implementation is area intensive. Hence, an effective search of the solution space yields solutions that exhibit tremendous trade-offs between delay and area. Thus a considerable reduction in average area is observed at the expense of delay. Thus, it can be concluded that NSGA-II, being a true Multi-objective GA is highly effective in more intensive exploration of the design space. This is evident from the quality metrics as well as the results on the various DFG benchmarks.

Table 5.5 lists out the comparison of the execution times. The NSGA-II algorithm shows appreciable improvement over WSGA with an average reduction of 51.63 % in execution time. As expected, the improvement is more noticeable in the case of the larger benchmarks which necessitate exploration of a much larger search space. The search space is multi-dimensional and non-linear and hence, increase in number of DFG nodes will exponentially add large complexity to the search process.

5.5.3 WSPSO – Results and Analysis

The WSPSO-based weighted sum approach outlined in Sect. 5.4.7 is evaluated on the DFG benchmarks and compared with WSGA. A particle size of 100 is used and the WSPSO is run for 70 generations. The average power, area and delay values of 10 runs of the algorithm with different initial seed populations is tabulated for various DFG benchmarks. The quality of the solutions is also assessed against the results from exhaustive search. The results are shown in Tables 5.6 and 5.7. Initial results are comparable to the corresponding results for WSGA. However, the quality and run times can be improved upon optimizing the swarm size and introducing additional operators in the velocity function used. Further investigations need to be carried out in this direction.

Tables 5.6 and 5.7 present the comparison of WSGA and WSPSO on standard DFG benchmarks whereas Table 5.6 compares average values of power, area, delay, and execution times and Table 5.7 lists out the performance metrics (Deb 2008) that indicate the quality of the solutions obtained in terms of diversity and closeness to the true Pareto front. This is assessed by generating the true Pareto Front using exhaustive search and comparing the Rank I solutions obtained from the algorithm with the True Pareto front. Low values for each metric indicate a higher degree of diversity and closeness to the Pareto front.

From Table 5.6, it can be observed that the results for WSPSO are comparable with WSGA for the area, delay, and power values whereas WSPSO exhibits an average improvement of 26.19 % in the execution times. Thus, there is scope for improvement in the WSPSO methodology by enhancing the search method without an adverse impact on execution times. However, WSPSO as in the case of WSGA, fares poorer than NSGA-II in terms of solution spread, diversity, closeness to Pareto Front as well execution times.

Average measurements are not indicative of the ability of the algorithm to search the entire multi-objective solution space. The diversity of solutions and optimality in terms of closeness to the true Pareto front are also indicators of algorithm

Table 5.6 Comparison of WSGA and WSPSO on DFG benchmarks

	WSGA				WSPSO				
Bench-Mark	Power	Area (register units)	Delay (time steps)	Execution time (seconds)	Power	Area (register units)	Delay (time steps)	Execution time (seconds)	Reduction in execution times %
IIR	0.9019	48.48	5.148	6	0.9083	35.93	5.86	4	33.33
HAL	0.9901	37.81	4.981	11	0.9960	33.75	4.98	8	27.27
DWT	0.7126	50.92	10	26	0.7242	26.57	10.11	14	46.15
FIR	0.6416	54.48	11.32	112	0.6595	44.68	10.59	97	13.39
MPEG	0.6714	107.5	9	48	0.6984	95.30	7.69	34	29.17
DCT	0.6086	79.06	12.18	1,528	0.6120	75.98	11.30	1,408	7.85

Table 5.7 Metrics evaluating closeness to the Pareto-optimal front for IIR benchmark for WSGA and WSPSO

Performance metrics	WSGA	WSPSO
Error ratio	1	1
Generational distance	2.09	3.02
Maximum Pareto optimal front error (MFE)	81	43.02
Spacing	2.73	0
Spread	1.56	1
Weighted metric	1.82	2.04

efficiency. The performance metrics (Deb 2008) are evaluated for WSGA as well as WSPSO are listed in Table 5.7. The error ratio is the same for both algorithms. WSPSO fares slightly better in yielding uniformly spaced solutions with better diversity as seen by the lower values of *spacing* and *spread* for WSPSO. However, the *generational distance* metric is higher for WSPSO. Thus, a higher *generational distance* metric coupled with a lower *MFE* indicates that Rank I individuals have more or less uniform separation from the True Pareto Front. This is corroborated by the lower values for spacing and spread mentioned earlier.

5.6 Conclusion

A framework for applying multi-objective evolutionary techniques for multi-objective power, area, and delay optimization for the datapath scheduling problem in High Level Synthesis is presented. The methodology has been applied to the NSGA-II algorithm and considerable improvement in runtimes and solution quality is demonstrated compared to a weighted sum GA approach by evaluating the techniques on standard DFG benchmarks. Thus, the technique will be effective as a tool for design space exploration during DFG synthesis. The methodology has been extended to weighted sum PSO and preliminary results indicate that WSPSO is faster than WSGA with the potential for improvement in solution convergence and quality. The following conclusions can be drawn from the analysis of the WSGA, NSGA-II, and WSPSO algorithms.

- NSGA-II exhibits the best solution quality among all the techniques analyzed and fastest execution times and emerges as the best approach among the three methods compared from the point of view of efficient design space exploration. This is in line with expectations since NSGA-II incorporates features for ensuring diversity and convergence to the Pareto front. The method for identifying non-dominated solutions is computationally fast thus resulting in the reduced run time.

- WSPSO is computationally simple and exhibits lesser run time compared to WSGA since the evolution process does not involve crossover and mutation which are computation intensive.
- However, WSPSO may not be an alternative to NSGA-II unless the solution quality is improved by incorporating some hybrid approaches like interweaving with Simulated Annealing as reported in (Fang et al. 2007). This aspect needs to be investigated and can be part of the future work.
- True Multi-objective PSO (MOPSO) techniques have been proposed such as the works reported in (Zitzler and Thiele 1998) and (Padhye et al. 2009). This can be adapted for the HLS scheduling problem and the performance vis-à-vis NSGA-II can be studied.
- The solutions obtained by both WSPSO, WSGA, and NSGA-II will be synthesized to a target library and the trends predicted by the DFG metrics used in the cost functions with actual estimates from post binding synthesis results can be compared.

Acknowledgements Sections 5.4, 5.5, and 5.6 are reproduced from the reference D. S. Harish Ram et al. (2012) under the creative commons attribution license from M/s Hindawi Publishing Corporation. This is gratefully acknowledged.

References

Abdel-Kader RF (2008) Particle swarm optimization for constrained instruction scheduling. VLSI Des 7(4):1–7

Casseau E, Le Gal B (2009) High-level synthesis for the design of FPGA-based signal processing systems. In: Proceedings of the ninth international symposium on systems, architectures, modeling and simulation, Greece, pp 25–32

Chabini N, Wolf W (2005) Unification of scheduling, binding and retiming to reduce power consumption under timing and resource constraints. IEEE Trans VLSI 13(10):1113–1126

Chang JM, Pedram M (1995) Register allocation and binding for low power In: Proceedings of the thirty second annual ACM/IEEE design automation conference, USA, pp 29–35

Chen D, Cong J, Fan Y, Wan L (2010) LOPASS: a low-power architectural synthesis system for FPGAs with interconnect estimation and optimization. IEEE Trans Very Large Scale Integr (VLSI) Syst 18(4):564–577

Deb K (2008) Multi-objective optimization using evolutionary algorithms. Wiley, Chichester

Deb K, Pratap A, Agarwal S, Meyarivan T (2002) A fast and elitist multi-objective Genetic Algorithm: NSGA-II. IEEE Trans Evol Comput 6(2):182–197

Fang L, Chen P, Liu S (2007) Particle swarm optimization with simulated annealing for TSP. In: Proceedings of sixth WSEAS international conference on artificial intelligence, Knowledge Engineering and Data Bases, Corfu, Greece, pp 206–210

Ferrandi F, Lanzi PL, Loiacono D, Pilato C, Sciuto D (2007) A multi-objective genetic algorithm for design space exploration in High-Level Synthesis. In: Proceedings of the sixth IEEE computer society annual symposium on VLSI, Porto Alegre, Brazil, pp 417–422

Gerez SH (2000) Algorithms for VLSI design automation. Wiley, USA

Harish Ram DS, Bhuvaneswari MC, Prabhu SS (2012) A novel framework for applying multi-objective GA and PSO based approaches for simultaneous area, delay, and power optimization in high level synthesis of datapaths. VLSI Design, vol 2012, Article ID 273276, 12 pages, 2012. doi:10.1155/2012/273276

Hashemi SA, Nowrouzian B (2012) A novel particle swarm optimization for high-level synthesis of digital filters. In: Proceedings of the twenty fifth IEEE international symposium on circuits and systems, pp 580–583

Jie J, Ji H, Wang M, Zhao M (2010) Improved discrete particle swarm optimization based on edge coding and multilevel reduction strategy for larger scale TSP. In: Proceedings of the sixth IEEE international conference on natural computation, Yantai, China, pp 2633–2637

Krishnan V, Katkoori S (2006) A genetic algorithm for the design space exploration of datapaths during high-level synthesis. IEEE Trans Evol Comput 10(3):213–229

Kursun E, Mukherjee R, Memik SO (2005) Early quality assessment for low power behavioral synthesis. J Low Power Electron 1(3):273–285

Lin D, Qiu S, Wang D (2008) Particle swarm optimization based on neighborhood encoding for traveling salesman problem. In: Proceedings of the IEEE international conference on systems, man and cybernetics, pp 1276–1279

Martin G, Smith G (2009) High-level synthesis: past, present, and future. IEEE Des Test Comput 26(4):18–25

Murugavel AK, Ranganathan N (2003) A game theoretic approach for power optimization during behavioral synthesis. IEEE Trans VLSI 1(6):1031–1043

Padhye N, Branke J, Mostaghim S (2009) Empirical comparison of MOPSO methods - guide selection and diversity preservation. In: Proceedings of the eleventh IEEE congress on evolutionary computation, pp 2516–2523

Paulin P, Knight JP (1989) Force-directed scheduling for the behavioral synthesis of ASICs. IEEE Trans Comput-Aided Des Integr Circuits Syst 8(6):661–679

Pilato C, Loiacono D, Tumeo A, Ferrandi F, Lanzi PL, Sciuto D (2010) Speeding-up expensive evaluations in high-level synthesis using solution modeling and fitness inheritance. In: Yoel Tenne, Chi-Keong Goh (eds) Computational intelligence in expensive optimization problems, vol 2. Springer, Berlin/Heidelberg, pp 701–723

Schafer BC, Wakabayashi K (2011) Divide and conquer high level synthesis design space exploration. ACM Trans Embed Comput Syst 9(4):39.1–39.17

Schafer BC, Wakabayashi K (2012) Machine learning predictive modelling high-level synthesis design space exploration. IET Comput Digital Techn 6(3):153–159

Sengupta A, Mishra VK (2014) Automated exploration of datapath and unrolling factor during power–performance tradeoff in architectural synthesis using multi-dimensional PSO algorithm. Elsevier J Expert Syst Appl 41(10):4691–4703

Sengupta A, Sedaghat R (2011) Integrated scheduling, allocation and binding in high level synthesis using multi structure Genetic Algorithm based design space exploration. In: Proceedings of the twelfth international symposium on quality electronic design, Santa Clara, CA, USA, pp 1–9

Wang KP, Huang L, Zhou CG, Pang W (2003) Particle swarm optimization for traveling salesman problem. In: Proceedings of the international conference on machine learning and cybernetics, pp 1583–1585

Xydis S, Palermo G, Zaccaria V, Silvano C (2013) A meta-model assisted coprocessor synthesis framework for compiler/architecture parameters customization. In: Proceedings of the sixteenth EDA consortium conference on design, automation and test in Europe, pp 659–664

Zitzler E, Thiele L (1998) Multi-objective optimization using evolutionary algorithms – a comparative case study. In: Parallel Problem Solving from Nature, Springer

Zuluaga M, Krause A, Milder P, Püschel M (2012) Smart design space sampling to predict Pareto-optimal solutions In: Proceedings of the thirteenth ACM SIGPLAN/SIGBED international conference on languages, compilers, tools and theory for embedded systems, Beijing, China, pp 119–128

Chapter 6
Design Space Exploration of Datapath (Architecture) in High-Level Synthesis for Computation Intensive Applications

Anirban Sengupta

Abstract Hardware accelerators (or custom hardware circuit) incorporate design practices that involve multiple convoluted orthogonal optimization requisites at various abstraction levels. The convoluted optimization requisites often demand intelligent decision-making strategies during high-level synthesis (HLS) to determine the architectural solution based on conflicting metrics such as power and performance as well as exploration speed and quality of results. Traditional heuristic-driven approaches using genetic algorithm, simulated annealing, etc., fall short considerably on the above orthogonal aspects especially in their ability to reach real optimal solution at an accelerated tempo. This chapter introduces a new particle swarm optimization-driven multi-objective design space exploration methodology based on power-performance trade-off tailored for targeting application-specific processors (hardware accelerators). Furthermore, as the performance of particle swarm optimization is known for being highly dependent on its parametric variables, in the proposed methodology, sensitivity analysis has been executed to tune the baseline parametric setting before performing the actual exploration process.

Keywords Design space exploration • High-level synthesis • Hardware accelerators • Particle swarm optimization • Sensitivity analysis

6.1 Introduction

As the design complexity increases in leaps and bounds by addition of more silicon per unit area, the design method for modular systems with multiparametric optimization objectives must be formalized. Simultaneously, optimizing multiple

A. Sengupta (✉)
Computer Science & Engineering, Indian Institute of Technology, Indore, India
e-mail: asengupt@iiti.ac.in

M.C. Bhuvaneswari (ed.), *Application of Evolutionary Algorithms for Multi-objective Optimization in VLSI and Embedded Systems*, DOI 10.1007/978-81-322-1958-3_6,

performance metrics with conflicting objectives such as power consumed and time of execution is becoming more complex and significant for successful design of these systems. Design decisions at the Electronic System Level (ESL) have more impact on the design quality than the decisions made at low level, i.e., logic level. For superior design quality, the assessment and selection should be comprehensive and accurate (Gajski et al. 1992).

Furthermore, the prominent increase in demand for portable embedded computing devices has led to new architecture exploration methodologies in high-level synthesis (HLS) intended for the design of hardware accelerators. The process of high-level synthesis is very complicated and descriptive and is traditionally performed by system architects. Depending on the application, the process of defining the problem, performing design space exploration (DSE), and the other required steps are time consuming. It should also be more efficient than the traditional methods in terms of time required to explore the design space and to select an optimum architecture that meets the multiple performance metric and constraints. However, despite some recent advancements in this area, the techniques proposed fall short in proposing a comprehensive universal and concurrent solution to twofold orthogonal problems encountered during DSE, namely, striking an efficient balance between conflicting metrics (power and performance) and providing a fast exploration process (at reduced time) that maximizes the quality of results (Mishra and Sengupta 2014).

6.2 Prior Works on Evolutionary-Algorithm-Based DSE

In Gallagher et al. (2004), evolutionary algorithm such as genetic algorithm (GA) has been proposed to yield better results for the DSE process. The use of GA has also been proposed in Krishnan and Katkoori (2006) as a promising framework for DSE of datapaths in high-level synthesis. Their work employs robust search capacities of the GA to resolve datapath synthesis of scheduling and allocation of resources. The fitness function in the above approaches (Gallagher et al. 2004) does not optimize total execution time (as a function of latency, initiation interval, and pipelined data sets). In Sengupta et al. (2012), the authors used multistructure chromosome representation for the datapath nodes for scheduling. The authors did not consider dynamic power while calculating total power. Another method introduced by researchers in Ferrandi et al. (2008) is based on solution encoding using genetic algorithm. The work optimizes area-latency during exploration and works for multimodal functional units also. However, the approach does not have any consideration toward execution time and power (which are very important design metrics). Besides, in Torbey and Knight (1998a, b), the authors showed another approach for DSE in HLS based on binary encoding of the chromosomes, which does not include execution time as design metric.

6.3 Multi-objective Design Space Exploration Using Particle Swarm Optimization

This chapter presents a design space exploration methodology called MO-PSE based on the particle swarm optimization (PSO) for designing application-specific processors. To the best of the authors' knowledge, this is the only work that directly transforms an independent and complete PSO process into a multi-objective DSE for power–performance trade-off of application-specific processors. Therefore, the major highlights of the work are as follows: it presents a PSO-driven DSE methodology for multi-objective trade-off, a fitness function used for assessment of cost each solution found, a mutation algorithm for diversity introduction thereby assisting in the improvement of convergence and exploration speed, an algorithm to handle boundary violation problem during exploration, and reports results based on the sensitivity analysis of PSO parameters such as swarm size, inertia weight, acceleration coefficient, and termination condition to assess its impact on the efficiency of the DSE.

The following subsections of this chapter describe in detail the above DSE process. Therefore, for the sake of clarity, the abbreviation and symbols used in the rest of the chapter are given in Fig. 6.1.

MO-PSE	- Multi objective particle swarm exploration
DSE	- Design space exploration
HLS	- High level synthesis
PSO	- Particle swarm optimization
SoC	- System on chip
L	- Lower boundary of design space
U	- Upper boundary of design space
R	- Resource configuration
R_{id}	- Resource configuration (position) of ith particle d^{th} dimension
R_i	- Resource configuration (position) of ith particle
R^+_{id}	- New Resource configuration (position) of ith particle d^{th} dimension
R_{lbi}	- Local best resource configuration of ith particle
R_{gb}	- Global Best Resource configuration
R_{lb}	- Local best Resource configuration
V	- Velocity
V_{id}	- Velocity of ith particle d^{th} dimension
V^+_{id}	- New velocity of ith particle d^{th} dimension
C_f	- Fitness Function

C_f^{lbi}	- Local best fitness of ith particle
C_f^{gb}	- Global best fitness among all particle
M	- Number of Iteration
T	- Current iteration
Ns	- Number of particle
D	- Number of dimension
d	- Current dimension
P_{min}	- Minimum power consumed by resource configuration
P_{cons}	- Power consumption constrained specified by user
P_{max}	- Maximum power consumed by resource c configuration
P_T	- Total Power consumed by resource configuration
T_{min}	- Minimum execution time taken by a resource configuration
T_{cons}	- Execution time constrained specified by user
T_{max}	- Maximum execution time taken by a resource configuration
T_E	- Execution time taken by a resource configuration
Z	- Stopping criterion

Fig. 6.1 Nomenclature for this chapter (*Note*: Reprinted from Vol 67/C, Mishra and Sengupta (2014), Copyright (2014), with permission from Elsevier)

6.3.1 Background on PSO

In PSO, each particle possesses a position and a velocity, which is changed as the particle flies around the multidimensional design space. As a derivative of the original velocity and position equation proposed in Kennedy and Eberhart (1995), in the proposed methodology, the resource position of i-th particle is changed by adding the velocity to the current position from Mishra and Sengupta (2014):

$$\begin{aligned}\text{Current position } (x_i) \text{ in time } (t+1) = {} & \text{Past position } (x_i) \text{ in time } (t) \\ & + \text{velocity}_i \text{ in time } (t+1)\end{aligned} \tag{6.1}$$

The velocity is updated with the following rule:

$$\begin{aligned}\text{Velocity}_i \text{ in time } (t+1) = {} & [\text{inertial weight } (w) * \text{velocity } (v)_i \text{ in time } (t)] \\ & + [\text{acceleration coefficient } (b_1) * \text{rand}_1\{\text{local best } (x_i{}^{\text{lb}}) \\ & - \text{current position } (x_i) \text{ in time } (t)\}] \\ & + [\text{acceleration coefficient } (b_2) * \text{rand}_2\{\text{global best } (x^{\text{gb}}) \\ & - \text{current position } (x_i) \text{ in time } (t)\}]\end{aligned} \tag{6.2}$$

where w is the inertia weight; b_1 is the cognitive learning factor; b_2 is the social learning factor; r_1, r_2 are random numbers in the range [0, 1]; x_i^{lb} is the best position of i-th particle; x^{gb} is the global best position found so far.

6.3.2 Generic Overview of MO-PSE

The block diagram of the proposed DSE methodology using PSO is shown in Fig. 6.2, which is called the multi-objective particle swarm exploration (MO-PSE) algorithm. Moreover, the flow chart containing the details of the block diagram is shown in Fig. 6.3. Its corresponding description is provided below.

The inputs to the framework are description of data flow graph (DFG) that describes datapath, user-specified design constraints for power and execution time (with user-specified weight factor) and library information. The framework also has the capability to check whether the user constraints are valid. In the framework presented, an independent PSO algorithm has been directly

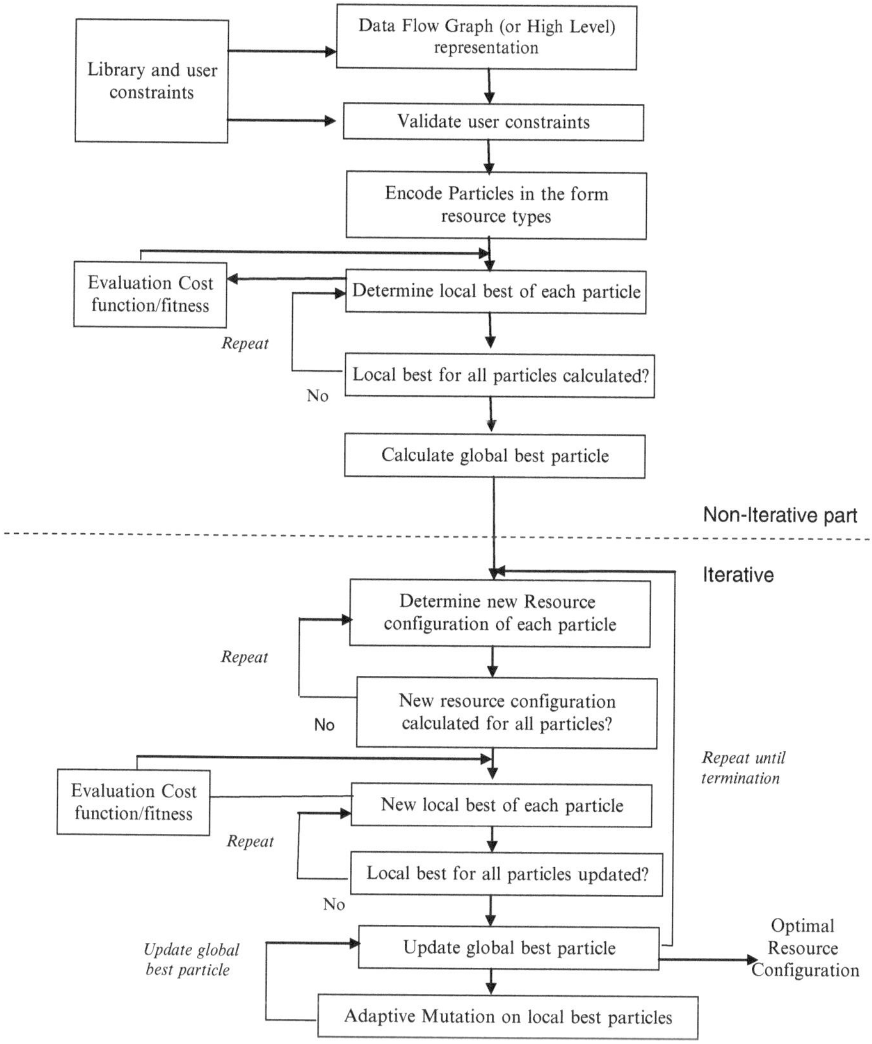

Fig. 6.2 Block diagram of the MO-PSE

transformed into the DSE process by the following mapping (Mishra and Sengupta 2014):

Position of particle → Resource configuration
Velocity of particle → Exploration deviation/drift
Dimension → # of Resource type

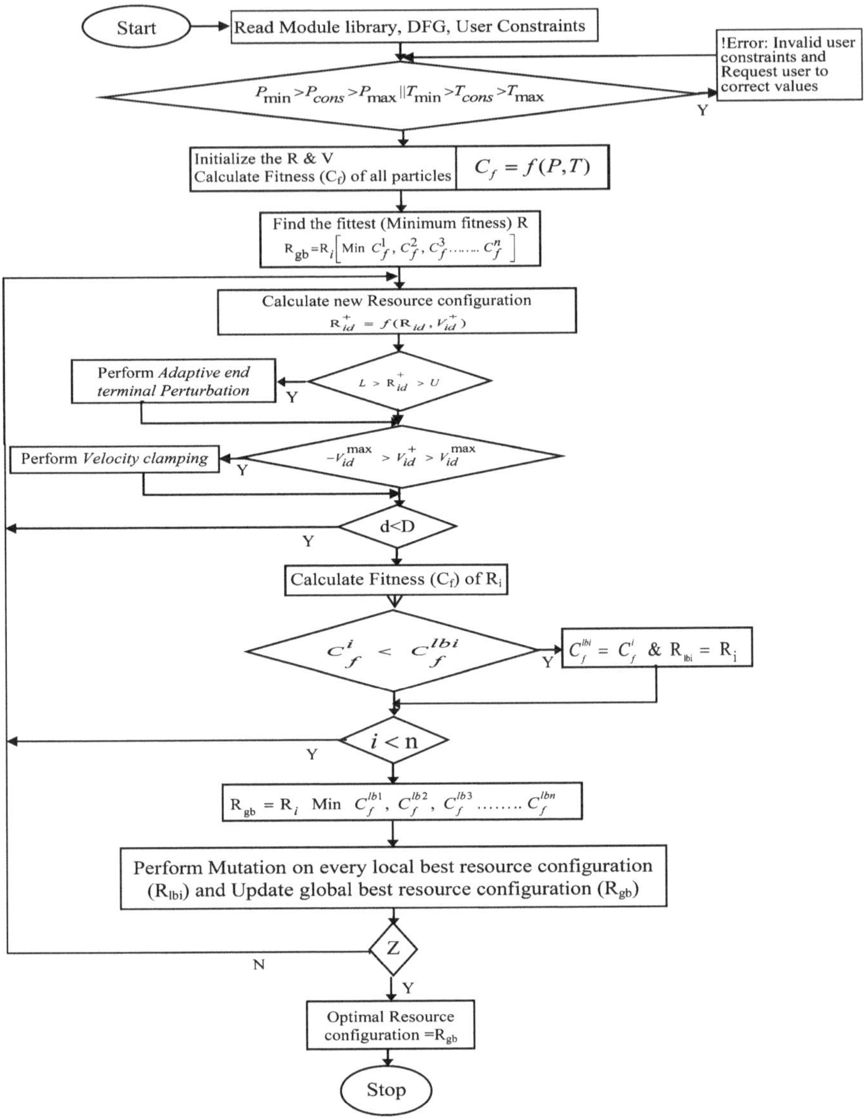

Fig. 6.3 Flow chart of proposed MO-PSE (*Note*: Reprinted from Vol 67/C, Mishra and Sengupta (2014), Copyright (2014), with permission from Elsevier)

6.3.3 Encoding of Swarms

Further, the swarm population is mapped as a set of initial design points in design space (considered as initial design solutions that will be subjected to improvement in each iteration), while the social and cognitive components of PSO are used as factors that supplement in exploration drift process during architecture optimization. The framework presented has multiple particles as initial swarm population. The first particle $P1$: (first initial design point) is encoded by mapping the minimum value resource configuration (Mishra and Sengupta 2014):

$$P1 = \left(R1^{min}, R2^{min}, R3^{min}.....Rn^{min}\right) \tag{6.3}$$

The second particle: $P2$ (second initial design point) is encoded by mapping the maximum value resource configuration (Mishra and Sengupta 2014):

$$P2 = \left(R1^{max}, R2^{max}, R3^{max}....Rn^{max}\right) \tag{6.4}$$

The third particle: $P3$ is encoded by mapping the average value of minimum and maximum resource configuration (Mishra and Sengupta 2014):

$$P3{=}\left(\left(R1^{min} + R1^{max}\right)/2, \left(R2^{min} + R2^{max}\right)/2, \left(R3^{min} + R3^{max}\right)/2\ldots\ldots \left(Rn^{min} + Rn^{max}\right)/2\right) \tag{6.5}$$

The rest of the particles ($P4....Pn$) are constructed as follows (Mishra and Sengupta 2014):

$$R_{id} = (a + b)/2 \pm \alpha \tag{6.6}$$

where "a" is minimum resource value, "b" is maximum resource value, and "α" is a random value between "a" and "b." Additionally in the proposed algorithm, all the particles' initial velocity is assumed to be zero. Once the particles are initialized, determination of particles fitness is performed and the fittest particle (which has minimum fitness) is chosen. The fittest particle becomes the global best resource configuration and fitness of this particle acts as best fitness for the next iteration. After this step, iteration process initiates. According to the algorithm, in each iteration, the new resource configurations (design points) of all particles based on the given function are upgraded and evaluated based on the following equation:

$$R_{id}^{+} = R_{id} + V_{id}^{+} \tag{6.7}$$

Equation 6.7 denotes the procedure for determining the next resource configuration during PSO using the information of the current position or resource configuration (R_{id}) and new velocity or exploration drift (V_{id}^{+}). The detailed description of Eq. 6.7 is given in this chapter. The process of determining the new resource configuration (R_{id}^{+}) needs critical discussion. Two cases may arise which lead to violations that are administered as follows:

(a) *Violation during resource up gradation:* Combated through "*adaptive end terminal perturbation*" algorithm redeeming any violations encountered during the DSE process. Pseudo-code is shown in Fig. 6.4.

```
Input - Resource configuration which crosses the design space
Output - New value of resource configuration with in design space

//When R_id crosses the design space boundary

While (R_id < L)
{
        R_id = R_id + Y
}
While (R_id > U)
{
        R_id = R_id - Y
}
  /* where 'Y' is a random value between minimum resource constraints and maximum resource
  constraints.
  'L' is lower boundary which means minimum resource value single instance.
  'U' is the upper boundary which means maximum # of resources*/
```

Fig. 6.4 Adaptive end terminal perturbation (*Note*: Reprinted from Vol 67/C, Mishra and Sengupta (2014), Copyright (2014), with permission from Elsevier)

(b) *Violation during velocity calculation:* Combated through "*velocity clamping*" in the MO-PSE thereby controlling excessive exploration drift. For more details, (Mishra and Sengupta 2014) can be referred.

After recalculation of the fitness of all particles with new resource configuration, updating the local best positions (resource configuration) and global best position (resource configuration) is performed when the new fitness is lower than the previous fitness. Once the global resource configuration is found, the mutation scheme as shown in Fig. 6.5 is applied on all local best resource configurations to explore a better global best resource configuration than the existing one. The aforementioned steps are repeated for next iterations until the stopping criterion is reached (the stopping criteria are described in this chapter later). Therefore, after completing the process, the algorithm yields the optimal architecture for a given DFG and user constraint.

6.3.4 *Mutation Algorithm in MO-PSE*

Adaptive rotation mutation algorithm as described in Mishra and Sengupta (2014) is an algorithm for mutation operation where the algorithm uses two basic operations for mutation. First is rotation operation, and second is increment or decrement operation. To perform these operations, the total population is divided into two groups: one is the even group in which algorithm performs left rotation operation and the second group is the odd group in which algorithm performs increment or decrement with a random number. After mutation algorithm calculates new fitness value of local best resource configuration, if the fitness is better, then the new value

```
Input – Local best resource configuration Rlb
Output – New mutated local best resource configuration Rlb

For i=1 to n // where n=Ns is the swarm size (# of particles)
{
        If (i%2==0) // Left Rotation
        {
                for j=1 to D
                {
                        Temp = Rj
                        Rj= Rj+1
                        Rj+1= temp
                        J++
                }
        }

        If(i%2==1)
        {
                For j=1 to D
                {
                        Rj = Rj ± X
                // X is a random number between [1,3]
                J++
                }
        }
        i++;
}
```

Fig. 6.5 Adaptive rotation mutation algorithm (*Note*: Reprinted from Vol 67/C, Mishra and Sengupta (2014), Copyright (2014), with permission from Elsevier)

will become the local best fitness, otherwise the older value does not change. The pseudo-code is shown in Fig. 6.5.

6.3.5 *Model for Evaluation of Particles (Design Points) During MO-PSE*

The fitness function (C_f), which considers total execution time and total power consumptions, is shown in Eq. (6.8) from Mishra and Sengupta (2014):

$$C_f = \varphi_1 \frac{P_{\mathrm{T}} - P_{\mathrm{cons}}}{P_{\max}} + \varphi_2 \frac{T_{\mathrm{E}} - T_{\mathrm{cons}}}{T_{\max}} \tag{6.8}$$

where C_f = Fitness of particle and the variables P_{T}, P_{cons}, $P_{\max}$, T_{E}, T_{cons}, $T_{\max}$ have been defined earlier, φ_1, φ_2 = User-specified weight factor for power and execution time.

The functions to calculate P_{T} and T_{E} are adopted from Sengupta et al. (2013). Figure 6.6 shows the process adopted to extract the hardware information for cost calculation.

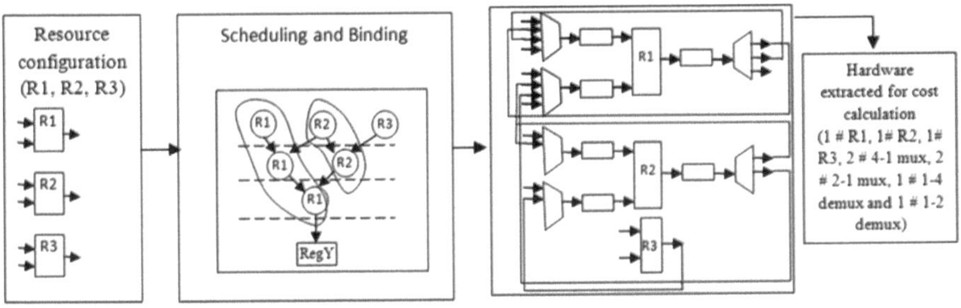

Fig. 6.6 Process of hardware information extraction during cost calculation

6.3.6 *Determination of Local and Global Best Positions/Resource Configurations*

As described in Mishra and Sengupta (2014), after calculating the fitness of each particle, the next step of the algorithm as shown in Fig. 6.3 is to determine the $R_{\mathrm{lb}i}$ of each particle and finally determine the global best particle (R_{gb}). For example, let us assume that the new fitness of a particle "R_{id}" is given as "C_f^i" and the fitness of the previous local best particle "$R_{\mathrm{lb}i}$" is given as "$C_f^{\mathrm{lb}i}$" Then, if the new fitness of particle is less than the current local best fitness, then R_i^+ becomes new $R_{\mathrm{lb}i}$ of the particle. Therefore, the process of determining upgraded $R_{\mathrm{lb}i}$ is as shown in Fig. 6.7. Since in iteration 1, there is no previous local best position ("$R_{\mathrm{lb}i}$"), therefore the current position (R_{id}) assumes the value of $R_{\mathrm{lb}i}$. Next, the R_{gb} of the population is determined using Eq. 6.9 as follows:

$$R_{\mathrm{gb}} = R_i\left[\mathrm{Min}\left(C_f^{\mathrm{lb}1}, C_f^{\mathrm{lb}2}, C_f^{\mathrm{lb}3} \ldots\ldots..C_f^{\mathrm{lb}n}\right)\right] \tag{6.9}$$

6.3.7 *Determination of New Configuration of the Particle (and Updated Local Best Particle)*

For the determination of new resource configuration of each particle i, each dimension (d) indicating the resource value (R_{id}) of the particle needs calculation as shown in Fig. 6.8. The R_{id}^+ is calculated using Eq. (6.4) as defined before. Further, V_{id}^+ is calculated using Eq. (6.10) from Mishra and Sengupta (2014):

$$V_{id}^+ = \omega V_{id} + b_1 r_1 \left[R_{\mathrm{lb}i} - R_{id}\right] + b_2 r_2 \left[R_{\mathrm{gb}} - R_{id}\right] \tag{6.10}$$

where, the aforementioned variables are defined in Fig. 6.1 (nomenclature).

```
Assume:
```
c_f^{lbi} - Local best Cost of ith particle

c_f^{i} - Fitness of ith particle

R_{lbi} – Local best resource configuration of ith particle

R_i – Resource configuration (position) of ith particle

if $(c_f^i < c_f^{lbi})$ *then,*

$c_f^{lbi} = c_f^i$

$R_{lbi} = R_i$

Fig. 6.7 Pseudo-code for local best up-gradation (*Note*: Reprinted from Vol 67/C, Mishra and Sengupta (2014), Copyright (2014), with permission from Elsevier)

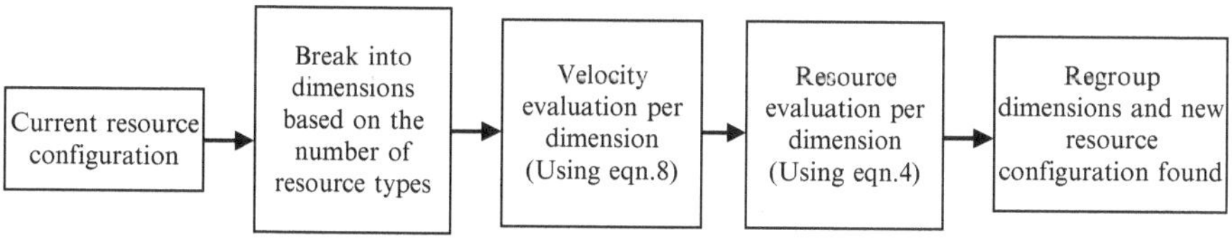

Fig. 6.8 Determination of new position of a particle

6.3.8 Stopping (Termination) Criterion (Z)

As described in Mishra and Sengupta (2014), the algorithm will terminate if one of following conditions holds true:

(a) Terminates when a maximum number of iteration has been exceeded ($M = 100$).
(b) Terminates when no improvement is observed over a certain number of iterations. We proposed two different stopping criteria as follows:
   1. S^1: When no improvement is seen in R_{gb} over "£" number of iteration (£ = 10).
   2. S^2: If the population reaches to equilibrium state, i.e., all particles velocity become zero ($V^+ = 0$).

The aforementioned conditions ensure that the algorithm does not prematurely converge and does not fall inside an indefinite loop.

6.4 Demonstration of MO-PSE with Examples

This section discusses the demonstration of the key steps of the MO-PSE algorithm, which includes encoding process of swarms, cost calculation, determination of new resource configuration, and mutation strategy with the help of a sample DFG of

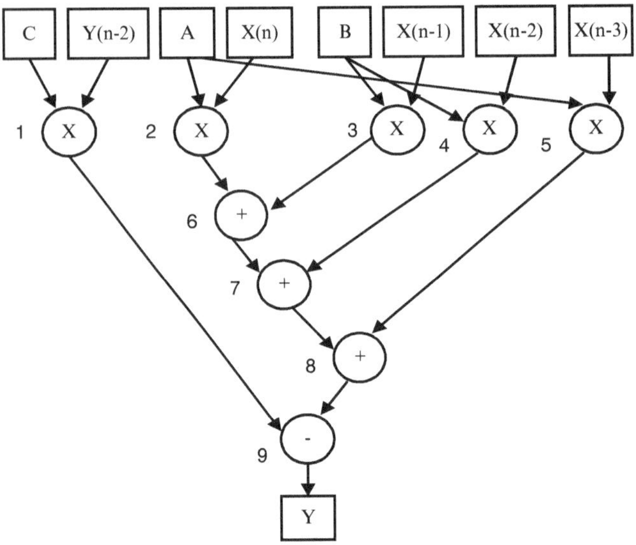

Fig. 6.9 Data flow graph of digital filter (*Note*: Reprinted from Vol 67/C, Mishra and Sengupta (2014), Copyright (2014), with permission from Elsevier)

digital filter. Some real values of module library information are assumed during demonstration of the steps. It is to be noted that the values chosen for constraints and filter DFG as application are purely arbitrary.

6.4.1 Encoding (Initialization) of Particles

Based on the definition of encoding of swarms (Eqs. 6.3, 6.4, 6.5, and 6.6), the particles encoded for the DFG (in Fig. 6.9) from Mishra and Sengupta (2014) are as follows:

$P1 = (1, 1, 1)$ {Based on the min value of resources}
$P2 = (6, 4, 4)$ {Based on the max value of resources}
$P3 = (3, 2, 2)$ {Based on the avg. value of resources}

6.4.2 Cost Calculation of Particles

Based on the cost function defined in Sect. 6.3.5 (Eq. 6.8), the fitness of particle in terms of arbitrary power and execution time constraints of 8 W and 310 us, respectively, is shown below (Mishra and Sengupta 2014):

Table 6.1 Example of determination of new configuration

Initial configuration of Particle $P1$	Dimensions (Particle: $P1$)	Velocity (V_{1d}^{+}) Eq. (6.8)	Resource (R_{1d}^{+}) Eq. (6.4)	New configuration of Particle $P1$
$R_1=(1, 1, 1)$	*'x' dimension*	$V_{1x}^{+}=1*0+2*0.5(1)+2*0.5(1)=2$	$R_{1x}^{+}=1+2=3$	$R_1^{+}=(3, 2, 2)$
	'y' dimension	$V_{1y}^{+}=1*0+2*0.5(1)+2*0.5(1)=1$	$R_{1y}^{+}=1+1=2$	
	'z' dimension	$V_{1z}^{+}=1*0+2*0.5(1)+2*0.5(1)=1$	$R_{1z}^{+}=1+1=2$	

For Particle ($P1$) with configuration (1, 1, 1):

$$C_f = 0.5\left(\frac{6.553 - 8}{31.159}\right) + 0.5\left(\frac{500.020 - 310}{500.020}\right)$$

$$C_f = 0.1667$$

where $P_{\mathrm{T}}=6.55$ W, $T_{\mathrm{E}}=500.02$ us, $P_{\mathrm{Max}}=31.15$ W, and $T_{\mathrm{Max}}=500.02$ us are evaluated from the models proposed in Sengupta et al. (2013) and library assumed in Mishra and Sengupta (2014). $\varphi_1=\varphi_1=0.5$ is chosen for providing equal weightage to power and execution time factor while calculating fitness of particle. Similarly, the fitness of all other particles is calculated using Eq. (6.6):

$C_f^{p1}=0.1667$ fitness of R_1 (1, 1, 1)
$C_f^{p2}=0.1691$ fitness of R_2 (6, 4, 4)
$C_f^{p3}=0.0116$ fitness of R_3 (3, 2, 2)

6.4.3 Determination of New Resource Configuration

Using Eqs. (6.4) and (6.8) defined earlier, the new resource configuration can be evaluated for particles in the design space as described in Mishra and Sengupta (2014). Table 6.1 shows the initial particle configuration of $P1$ and its corresponding new configuration. Similarly for other particles, the new resource configuration can be determined. In order to achieve convergence for the algorithm, it has been theoretically established before in Trelea (2003) that the cognitive learning factor (b_1) and the social learning factor (b_2) can be initialized to any value between (Mishra and Sengupta 2014). Therefore, $b=2$ has been used during evaluation.

6.4.4 Mutation

Based on the mutation algorithm defined in Sect. 6.3.4, this section provides examples of performing mutation on the local best positions of the three particles.

Table 6.2 Example of mutation

Particle local best	Before mutation	After mutation	Mutation operation
R_{lb1}	(3, 2, 2)	(4, 2, 3)	Increment/decrement by one
R_{lb2}	(2, 1, 1)	(1, 1, 2)	Rotation (left) operation
R_{lb3}	(4, 3, 3)	(3, 4, 2)	Increment/decrement by one

Table 6.2 shows an example of the mutation strategy as shown in Mishra and Sengupta (2014).

6.5 Experiments and Results

The results reported from Mishra and Sengupta (2014) in this chapter are based on the implementation carried in Java language on Intel core i5-2450M processor with 3MBL3 cache memory and 4 GB DDR3 primary memory at the processor frequency of 2.5 GHz.

The results are divided into two subsections.

6.5.1 *Sensitivity Analysis to Assess the Impact on MO-PSE Methodology*

(a) *Inertia weight (w):* The following three cases have been analyzed. (i) Linearly decreasing "*w*" in every iteration between [0.9–0.1] throughout the exploration process; (ii) a constant value of $w = 1$ throughout the exploration process; and (iii) a constant value of $w = 0.5$ throughout the exploration process.

In case (i), linearly decreasing *w* (with higher values of *w* initially and then gradually decreasing) is analyzed to equip the algorithm with higher exploration ability initially followed by higher exploitation ability. This technique assists the algorithm from escaping local minima and moving quickly out of nonoptimal architectures. On the contrary, in cases (ii) and (iii), fixed values of *w* are kept during the entire exploration process. The benchmark results reported in Mishra and Sengupta (2014) (in Table 6.3) suggest that the exploration time to reach an optimal solution for case (i) is lower than the other two cases. Therefore, case (i) is selected as a baseline parameter while performing the exploration process.

(b) *Acceleration Coefficients (b):* During experimentation "*b*1" and "*b*2" are kept equal to "*b*." Four different values of *b* are taken for experimental analysis in Mishra and Sengupta (2014), namely, $b = 1,2,3,4$. As evident from the results (Table 6.4) obtained in Mishra and Sengupta (2014), selecting a specific value of *b* always for any benchmark for yielding an optimal results is not trivial.

Table 6.3 Comparison of exploration time (ms) with respect to parameter "w"

	Linearly decreasing	w (0.5)	w (1)
IIR Butterworth	56.5	69.5	71.33
BPF	261	244	238
EWF	323	320	368
ARF	252.16	263	269.33
JPEG SAMPLE	205	206	252.16
MESA	78.5	86	109.33
FIR	136	147	122.66
MPEG	212	227	213.66

Note: Reprinted from Vol 67/C, Mishra and Sengupta (2014), Copyright (2014), with permission from Elsevier

Table 6.4 Comparison of exploration time (ms) with respect to parameter b

	b(1)	b(2)	b(3)	b(4)
IIR Butterworth	55.5	56.5	60.166	56.833
BPF	200	261	195.166	205.5
EWF	328.66	323	316.66	313.66
ARF	274	252	251.83	287
JPEG SAMPLE	207.33	205.16	228.33	205.15
MESA	76.833	78.5	79.66	82.83
FIR	149.33	136	147.16	153
MPEG	275.66	212.66	199	288

Note: Reprinted from Vol 67/C, Mishra and Sengupta (2014), Copyright (2014), with permission from Elsevier

However, it can be safely assumed that the best quality solutions for attaining faster convergence and exploration speed are clustered at $b = 2$ or $b = 3$ in most tested cases.

(c) *Stopping criterion (Z):* Based on the stopping criteria discussed in Sect. 6.3.8, an experiment is carried out to compare S^1 and S^2 in terms of convergence time taken to yield an optimal solution. As reported in Mishra and Sengupta (2014), criterion S^1 yields faster convergence to an optimal architecture for all the tested benchmarks. Therefore, S^1 is selected as a baseline parameter while performing the exploration process (Table 6.5).

6.5.2 Comparison with Current Evolutionary Approaches

The following parameters are reported in this section during comparison of MO-PSE with (Krishnan and Katkoori 2006): (a) Implementation run time, (b) Resource configuration, (c) Execution time, (d) Power, and (e) Quality of Results (QoR). The following settings are maintained based on the inferences drawn from the obtained results in Sect. 6.5.1: $\varphi 1$ and $\varphi 2$ equal to 0.5, the value of w is linearly decreased between [0.9–0.1], value of $b = 2$, swarm size = 3, stopping criterion = S^1 and $M = 100$.

Table 6.5 Comparison of convergence time (ms) with respect to stopping criterion ($S^1$and S^2)

Benchmarks	S^1	S^2
IIR Butterworth	23.5	42.66
BPF	121.66	171.166
EWF	116.166	233.833
ARF	127	984
JPEG SAMPLE	80.5	219
MESA	39	78.5
FIR	73.66	486.5
MPEG	100.5	390

Note: Reprinted from Vol 67/C, Mishra and Sengupta (2014), Copyright (2014), with permission from Elsevier

Results reported from Mishra and Sengupta (2014), as shown in Table 6.6 indicate that run time of MO-PSE is significantly superior compared to (Sengupta et al. 2012). Further, in terms of meeting user constraint, (Sengupta et al. 2012) suffers from violation of power for most benchmarks. On the other hand, no such violation is observed with MO-PSE. Finally, for all the benchmarks, the QoR (lower cost) is better for MO-PSE compared to (Sengupta et al. 2012). The underlined values in the table indicate violations.

Further, results of comparison with (Krishnan and Katkoori 2006), reported in Mishra and Sengupta (2014) (as shown in Table 6.7), indicate improvements in exploration speed (marginal though). However, significant improvement in QoR is obtained as compared to (Krishnan and Katkoori 2006) for all the benchmarks tested. Further, empirical evidence also reveals that the MO-PSE is able to avoid violations in execution time as well as power in almost all cases. On the contrary, power violations are noted for (Krishnan and Katkoori 2006), when tested for some benchmarks. Since MO-PSE is a PSO-based parallel evolutionary algorithm, where multiple particles participate in the exploration process; therefore, it assures escaping the local optima. Moreover, inclusion of proposed mutation strongly reduces any chance of local optimal convergence. Further, the solutions obtained for the tested benchmarks are real optimal solutions, which can be verified by comparing with the golden solutions found by exhaustive analysis.

6.6 Conclusion

This chapter presents a direct transformation of an independent and complete PSO process into a multi-objective DSE (MO-PSE) for power–performance trade-off of application-specific processors. The methodology incorporates a number of intelligent and adaptive decision-making strategies, which provide a foundation for quick exploration of an optimal architecture. MO-PSE is compared to current methodologies such as (Krishnan and Katkoori 2006) to adjudge the efficiency of the approach. Results indicated that MO-PSE is superior in terms of exploration speed and quality of results.

Table 6.6 Experimental result of comparison with Sengupta et al. (2012) for the tested benchmarks

Benchmark	Parameters of comparison											
	Implementation runtime		Resource configuration		Mux and demux		Execution time (us)		Power (W)		QoR (cost)	
	MO-PSE	Sengupta et al. (2012)	MO-PSE	Sengupta et al. (2012)	MO-PSE	Sengupta et al. (2012)	MO-PSE	Sengupta et al. (2012)	MO-PSE	Sengupta et al. (2012)	MO-PSE	Sengupta et al. (2012)
IIR filter	0.065 s	20s	2(*), 1 (+)	4(*), 1(+)	6 (mux), 3 (demux)	10(mux), 5 (demux)	220.01 us	120 us	7.01 W	11.9 W	0.43	0.48
							Constraint = *300 us*		*Constraint* = *8 W*			
MESA Horner Bezier	0.078 s	2.30 min	3(*), 1 (+)	3(*), 1(+)	8 (mux), 4 (demux)	8 (mux), 4 (demux)	100.1 us	100 us	9.61 W	9.65 W	0.36	0.36
							Constraint = 400 us		Constraint = 8 W			
ARF	0.252 s	10.5 min	3(*), 1 (+)	4(*), 1(+)	8 (mux), 4 (demux)	10 (mux), 5 (demux)	520.2 us	341 us	9.49 W	11.9 W	0.34	0.33
							Constraint = *600 us*		*Constraint* = *10 W*			
EWF	0.323 s	10.1 min	2(*), 1 (+)	2(*), 4(+)	4 (mux), 2 (demux)	12 (mux), 6 (demux)	320 us	200 us	7.02 W	13.1 W	0.47	0.63
							Constraint = *500 us*		*Constraint* = *8 W*			
FIR	0.136 s	3.20 min	3(*), 1 (+)	4(*), 2(+)	6 (mux), 3 (demux)	16 (mux), 8 (demux)	220.02 us	201 us	9.51 W	14.01 W	0.27	0.32
							Constraint = *500 us*		*Constraint* = *11 W*			
MPEG MMV	0.212 s	6.43 min	4(*), 1 (+)	5(*), 1(+)	10 (mux), 5 (demux)	12 (mux), 6 (demux)	340 us	281 us	11.9 W	14.4 W	0.25	0.26
							Constraint = *600 us*		*Constraint* = *12 W*			
BPF	0.261 s	4.33 min	2(*), 1 (+)	4(*), 2(+)	4 (mux), 2 (demux)	12 (mux), 6 (demux)	500.04 us	140 us	7.01 W	14.0 W	0.46	0.50
							Constraint = *600 us*		*Constraint* = *12 W*			
JPEG downsample	0.205 s	7.59 min	2(*), 4 (+)	1(*), 2(+)	12 (mux), 6 (demux)	6 (mux), 3 (demux)	140.32 us	300 us	13.13 W	6.57 W	0.36	0.37
							Constraint = *400 us*		*Constraint* = *10* W			

Note: Reprinted from Vol 67/C, Mishra and Sengupta (2014), Copyright (2014), with permission from Elsevier

Table 6.7 Experimental result of comparison with Krishnan and Katkoori (2006) for the tested benchmarks

	Parameters of comparison											
	Implementation runtime (sec)		Resource configuration		Mux and demux		Execution time (us)		Power (W)		QoR (cost)	
Benchmark	MO-PSE	Krishnan and Katkoori (2006)	MO-PSE	Krishnan and Katkoori (2006)	MO-PSE	Krishnan and Katkoori (2006)	MO-PSE	Krishnan and Katkoori (2006)	MO-PSE	Krishnan and Katkoori (2006)	MO-PSE	Krishnan and Katkoori (2006)
IIR filter	0.065 s	0.099 s	2	(*), 1(+) 9.42 W	3(*), 1(+)	6 (mux), 3 (demux)	8(mux),	4 (demux) 0.43	0.55	220.01 us	260 us	7.01 W
Constraint = 300 us		*Constraint = 8 W*										
MESA Horner Bezier	0.078 s	0.162 s	3(*), 1 (+)	2(*), 1(+)	8 (mux), 4 (demux)	6 (mux), 3 (demux)	100.1 us	620 us	9.61 W	6.95 W	0.36	0.60
							Constraint = 300 us		Constraint = 8 W			
ARF	0.252 s	0.343 s	3(*), 1 (+)	4(*), 1(+)	8 (mux), 4 (demux)	10 (mux), 5 (demux)	520.2 us	580 us	9.49 W	11.8 W	0.34	0.40
							Constraint = 600 us		*Constraint = 10 W*			
EWF	0.323 s	0.670 s	2(*), 1 (+)	1(*), 1(+)	6 (mux), 3 (demux)	4 (mux), 2 (demux)	320 us	1180 us	7.02 W	4.49 W	0.47	0.94
							Constraint = 500 us		*Constraint = 8 W*			
FIR	0.136 s	0.296 s	3(*), 1 (+)	2(*), 2(+)	8 (mux), 4 (demux)	8 (mux), 4 (demux)	220.02 us	540 us	9.51 W	8.98 W	0.27	0.48
							Constraint = 500 us		*Constraint = 11 W*			
MPEG MMV	0.12 s	0.312 s	4(*), 1 (+)	4(*), 2(+)	10 (mux), 5 (demux)	12 (mux), 6 (demux)	340 us	480 us	11.9 W	13.91 W	0.25	0.32
							Constraint = 600 us		*Constraint = 12 W*			
BPF	0.261 s	0.421 s	2(*), 1 (+)	2(*), 2(+)	4 (mux), 2 (demux)	8 (mux), 4 (demux)	500.04 us	600 us	7.01 W	8.98 W	0.46	0.57
							Constraint = 600 us		*Constraint = 12 W*			
JPEG downsample	0.205 s	0.546 s	2(*), 4 (+)	2(*), 1(+)	12 (mux), 6 (demux)	6 (mux), 3 (demux)	140.32 us	600 us	13.13 W	6.524 W	0.36	0.62
							Constraint = 400 us		*Constraint = 10* W			

Note: Reprinted from Vol 67/C, Mishra and Sengupta (2014), Copyright (2014), with permission from Elsevier

Acknowledgment I thank Elsevier for allowing the reuse of the materials from *'Elsevier Journal on Advances in Engineering Software'*, *Vol 67, Vipul Kumar Mishra, Anirban Sengupta, MO-PSE: Adaptive Multi Objective Particle Swarm Optimization Based Design Space Exploration in Architectural Synthesis for Application Specific Processor Design, Pages. 111–124,* for my publication of book chapter. I would also like to thank Indian Institute of Technology, Indore, India, for providing me with all the resources for carrying out this work.

References

Ferrandi F, Lanzi PL, Loiacono D, Pilato C, Sciuto D (2008) A multi-objective genetic algorithm for design space exploration in high-level synthesis. ISVLSI:417–422

Gajski D, Dutt ND, Wu A, Lin S (1992) High level synthesis: introduction to chip and system design. Kluwer Academic Publishers, Norwell

Gallagher JC, Vigraham S, Kramer G (2004) A family of compact genetic algorithms for intrinsic evolvable hardware. IEEE Trans Evolut Comput 8(2):1–126

Kennedy J, Eberhart RC (1995) Particle swarm optimization. In: Proceedings of the IEEE international conference on neural networks, Anchorage, pp 1942–1948

Krishnan V, Katkoori S (2006) A genetic algorithm for the design space exploration of datapaths during high-level synthesis. IEEE Trans Evolut Comput 10(3):213–229

Mishra VK, Sengupta A (2014) MO-PSE: adaptive multi objective particle swarm optimization based design space exploration in architectural synthesis for application specific processor design. Elsevier J Adv Eng Softw 67:111–124

Sengupta A, Sedaghat R, Sarkar P (2012) A multi structure genetic algorithm for integrated design space exploration of scheduling and allocation in high level synthesis for DSP kernels. Elsevier J Swarm Evolut Comput 7:35–46

Sengupta A, Mishra VK, Sarkar P (2013) Rapid search of pareto fronts using D-logic exploration during multi-objective tradeoff of computation intensive applications. In: Proceedings of IEEE 5th Asian Symposium on Quality Electronic Design (ASQED), Malaysia, pp 113–122

Torbey E, Knight J (1998a) High-level synthesis of digital circuits using genetic algorithms. In: Proceedings of the international conference on evolutionary computation, Anchorage, pp 224–229

Torbey E, Knight J (1998b) Performing scheduling and storage optimization simultaneously using genetic algorithms. In: Proceedings of the IEEE midwest symposium on circuits and systems, Anchorage, pp 284–287

Trelea CI (2003) The particle swarm optimization algorithm: convergence analysis and parameter selection. Elsevier Inf Process Lett 85(6):317–325

Chapter 7
Design Flow from Algorithm to RTL Using Evolutionary Exploration Approach

Anirban Sengupta

Abstract The design of modern Very Large Scale Integration (VLSI) devices from a higher abstraction level (algorithmic level) yields much greater productivity compared to designing at lower abstraction levels. However, designing from higher abstraction level (achieved through a process called high-level synthesis) signifies lack of lower-level details during parametric evaluation of alternative architectural choices. Therefore, an ideal solution is to perform meet-in-the middle methodology to reinforce the advantages of both top-down (from algorithmic level) and bottom-up design approaches. This chapter presents a formal design flow from algorithmic level to register transfer level using evolutionary approach as an exploration framework for hardware accelerators. The design process presented using evolutionary techniques is capable of directly converting an application (specified through a control data flow graph) from algorithmic level to its circuit structure at register transfer level.

Keywords VLSI • Algorithmic level • Evolutionary algorithm • Hardware accelerator • High level synthesis • RTL

7.1 Introduction

Very large Scale Integration (VLSI) designing can be broadly divided into multiple abstraction levels: algorithmic level, register transfer level (RTL), and layout level. Traditional design approach of digital systems have been mostly from RTL whereby the designer based on the user-defined objectives implements the application in the form of hardware description language or schematic capture. However, due to the explosion of various sophisticated devices with increased complexity, the

A. Sengupta (✉)
Computer Science & Engineering, Indian Institute of Technology, Indore, India
e-mail: asengupt@iiti.ac.in

M.C. Bhuvaneswari (ed.), *Application of Evolutionary Algorithms for Multi-objective Optimization in VLSI and Embedded Systems*, DOI 10.1007/978-81-322-1958-3_7,

development of a formal design methodology with concurrent satisfaction of multiple design metrics has become extremely significant. This formal design methodology should ideally integrate an efficient design space exploration (DSE) with the high-level synthesis (HLS) design steps (which reduces the manual effort of optimizing and designing a system at RTL, yielding greater productivity in terms of design time and quality). It is well-known that the problem of DSE is orthogonal and NP-hard in nature. Therefore, heuristics such as evolutionary algorithms (EAs) cater best to solve such an intricate problem. The element of randomness of EAs provides the diversity needed during the exploration process. Further, these techniques being parallel exploration approaches assure escaping local minima during searching the optimal architecture in the assorted design space. Owing to the aforementioned aspects, this chapter integrates and presents a formal design flow from algorithmic description of an application to RTL using EAs (Gajski et al. 1992).

7.2 Design Flow from Algorithmic Level to RTL Using EAs

This section presents the formal design flow from the algorithmic level to RTL using EAs. It is worthy to note that EA-based DSE can be driven by genetic algorithms (GAs) (Krishnan and Katkoori 2006) or particle swarm optimization (PSO) (Mishra and Sengupta 2014), or a hybrid genetic PSO technique (Harish Ram et al. 2012).

7.2.1 Problem Formulation and Defining Module Library

Problem formulation involves translating the application from a high-level description (mostly C/C++) into a data flow graph (DFG) or control data flow graph (CDFG). These graphs specify the input/output relation of the algorithm and the data dependency present in the data flow. The graph is defined in terms of its vertices and edges, where the vertices signify the operations and the edges indicate the data dependency present in the function (Torbey and Knight 1998).

Defining module library is also performed at the beginning of the design flow where information on components such as area, delay, and power are defined inside the library (Fig. 7.1).

7.2.2 Problem Encoding Using Chromosomes/Particles and Deciding Population Size

Problem encoding is a very important part of evolutionary algorithms. Suitable problem encoding is needed to assist in quick and efficient exploration. There have

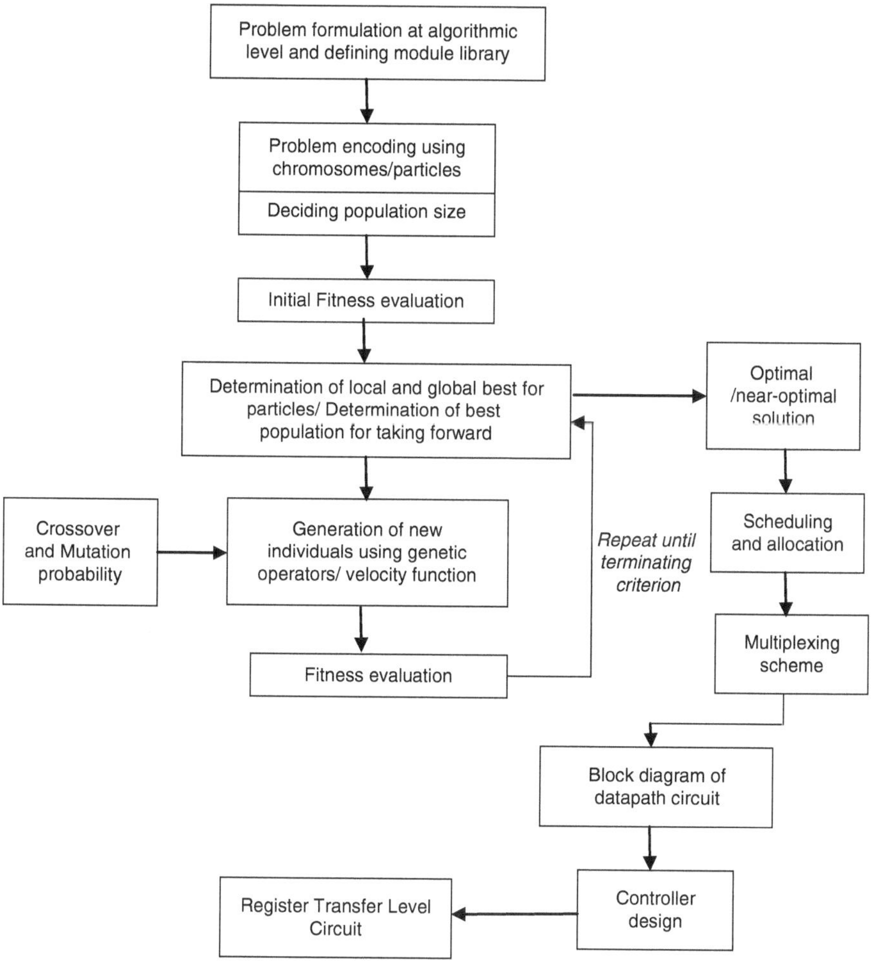

Fig. 7.1 Design flow from algorithmic level to RTL using EAs

been multiple types of problem encoding proposed for nodes (of a DFG) by researchers including (a) work remaining field of nodes (Dhodhi et al. 1995), (b) node priority field (Krishnan and Katkoori 2006), (c) load factor of nodes (Sengupta et al. 2012), and (d) number of successor nodes (Papa and Šilc 2002). However, there are other variances. Further, some approaches such as (Krishnan and Katkoori 2006) have proposed independent strings for encoding the nodes and resources (for allocation). This technique invokes the nodal string and resource allocation string independent of each other. Some works such as Mishra and Sengupta (2014) have presented multiple particle encoding based on different resource types being initialized in various dimensions. On the contrary, an independent PSO-based EA is encoded with dimensions represented as resource types (Mishra and Sengupta 2014).

Fig. 7.2 General problem encoding of nodes in EAs

Rn	E1	E2	E3	E4	E5	..	..	En
	Opn1	Opn2	Opn3	Opn4	Opn5	...	...	opn n

7.2.2.1 General Representation of Nodes in a Chromosome

The general problem encoding structure of nodes using GA is represented as follows (Fig. 7.2):

"En" is an encoded value of a node/operation (opn n). However, with respect to Dhodhi et al. (1995), "En" is encoded as follows:

$$\mathrm{En} = \begin{cases} \mathrm{WR} & \text{or,} \\ L & \text{or,} \\ S \end{cases}$$

where WR = work remaining field of each node obtained from DFG (Dhodhi et al. 1995); L = load factor of each node obtained from DFG (Sengupta et al. 2012); S = number of successor nodes obtained from DFG (Papa and Šilc 2002); Rn = operations of only a single resource type (such as multiplier/adder/sub) or can comprise operations of all resource types combined into a single chromosome.

7.2.2.2 Demonstration with an Example

A sample DFG shown in Fig. 7.3 has been used to demonstrate an example of encoding process based on Sengupta et al. (2012). It is to be noted that other approaches could also have been used for demonstration.

An encoded chromosome of resource type multiplier based on load factor "L" can be represented as shown in Fig. 7.4. It is to be noted that the delay of a single multiplier and adder/sub is assumed to be "$D+x$" cc and "D" cc, respectively. Therefore, as described in Sengupta et al. (2012), "L" value for node/opn 1 is the summation of all its dependent nodes including the opn 1 itself. Hence, L = "$5D+x$" cc. Other nodes of multiplier operation are also encoded in a similar way. Using this process, all the chromosomes of other resource types can be encoded.

In the representation in Fig. 7.4 and general structure in Fig. 7.2, the higher the encoded value (L, WR, or S), the greater is the preference of the node during scheduling. So, if more than two multiplication operations are competing for two available multiplier resources, then the top two multiplication operations with higher encoded value will be scheduled first while the rest are done later.

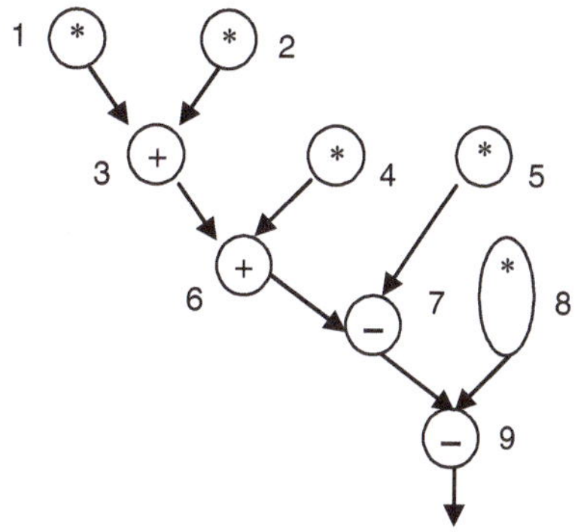

Fig. 7.3 Sample DFG

Fig. 7.4 Problem encoding of nodes using EAs (Dhodhi et al. 1995)

	5D +x	5D +x	3D +x	2D +x
Mul	1	2	5	8

a

I_{R1}	I_{R2}	I_{R3}	I_{Rn}
R1	R2	R3	Rn

b

1	2
A	M

Fig. 7.5 (**a**) General problem encoding of nodes using EAs. (**b**) Problem encoding of nodes in EAs using Sengupta et al. (2012)

7.2.2.3 General Representation of Resource Types in a Chromosome

The problem of encoding the resource types is often encoded in a string called "resource allocation string" (Sengupta et al. 2012). The encoded values of the resource allocation string consists of random integer values of corresponding resource types (without violating its maximum value) for each individual.

A general problem encoding structure of resources using genetic algorithms can be represented as in Fig. 7.5a, where, I_{Rn} = Random integer value of resource Rn without violating the Rn max (its corresponding maximum value) in an individual (parent or offspring).

7.2.2.4 Demonstration with an Example

Based on the DFG shown in Fig. 7.3, there are two types of resources adder/subtractor (A) and multiplier (M). Lets us assume adder/subtractor max = 4 and multiplier max = 5. Therefore, an individual can be randomly encoded in the resource allocation string as in Fig. 7.5b.

7.2.3 Fitness Evaluation and Determination of Best Population for Forward Generation

Cost/fitness evaluation of the population needs to be performed to identify how healthy each individual is. A general fitness function consisting of weighing objective metrics used in EA based DSE can be generically represented as shown in Eq. 7.1:

$$Cf = W1\frac{M1 - \text{constraint}\ (M1)}{M1\text{max}} + W2\frac{M2 - \text{constraint}\ (M2)}{M2\text{max}} + \cdots + Wn\frac{Mn - \text{constraint}\ (Mn)}{Mn\text{max}} \tag{7.1}$$

where,

"Wn" = weighing factor for metric "Mn" (usually assigned between 0 and 1);
"Mn" = parametric value of the metric (such as power, area, delay, and execution time);
"Mn max" = maximum value of metric Mn (used for normalizing the value of the metric);
"Cf" = cost of the architectural solution (usually lies between 0 and 1).
"constraint (Mn)" = user-specified constraint specified for metric "Mn."

After the fitness values of each individual is calculated, the local best and global best (in case of particle swarm) or set of best individuals (in case of GA) are forwarded to next generation/iteration. The process of calculating local and global best in case of particle swarm is described in Chap. 6 using Mishra and Sengupta (2014).

7.2.4 Generation of New Individuals Using Evolutionary Operators

In order to generate new individuals, evolutionary algorithms have special operators called crossover, mutation, and velocity function. These operators have various ways of implementation as described in Sengupta et al. (2012). Crossover in GAs is responsible for generating offsprings from parents while mutation is responsible for introducing randomness in the exploration process. On the other hand, generation of new individuals in case of PSO is performed using velocity function. The velocity value is then added to the current position of the particle to obtain a new position. The details of implementing the velocity function are described in Eqs. (1) and (2) in Chap. 6. However, the crossover technique employed in GA-based exploration is described as follows: Assume a parent (P1) and parent (P2) encoded with En through "WR," "*L*," or "*S*" as shown in Fig. 7.6. A crossover operation at cut point 2 yields an

P1 (Mul)	E1	E2	E3	E4
	op1	op2	op3	op4

P2 (Mul)	E5	E6	E7	E8
	op1	op2	op3	op4

Fig. 7.6 Sample problem encoding of nodes for two parents

Offspring	E1	E2	E7	E8
	op1	op2	op3	op4

Fig. 7.7 Offspring generated through two parents

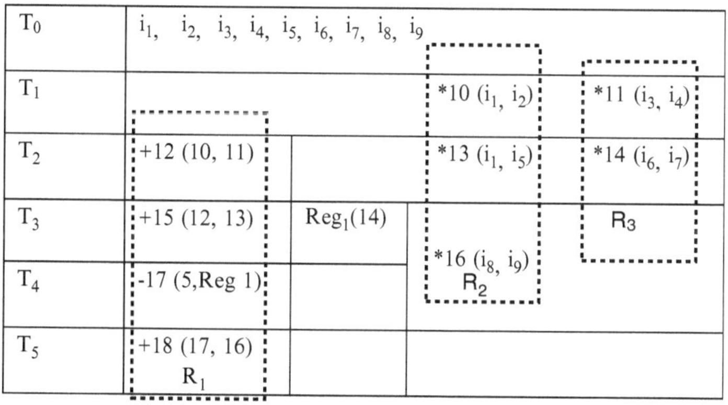

Fig. 7.8 Scheduling of the DFG

offspring as shown in Fig. 7.7. The offspring generated contains new combination of encoded values for the nodes that produce new scheduling solutions during exploration.

7.2.5 *Using an Optimal/Near-Optimal Solution Produced for Scheduling and Allocation*

The steps based on the flow chart shown in Fig. 7.1 discussed so far are iteratively executed until the terminating condition of the EA is reached, which in turn is able to produce an optimal/near-optimal architectural configuration (solution) for the application based on the multi-objective user constraints provided (Fig. 7.8).

Once the optimal/near-optimal solution is produced, it is subjected to final scheduling and allocation. For example, let us assume for demonstration, an individual with chromosomal information shown in Fig. 7.4 (nodal string) and Fig. 7.6 (resource string) for sample DFG (shown in Fig. 7.3) is yielded as the final architectural solution. This nodal string indicates that among the four independent operations opn1, 2, 3, and 4, opn 1 and 2 have the highest "*L*" value compared to opn 3 and 4, respectively. Therefore, opn 1 and 2 will be given priority

Algorithm

1. Create a table with 5 columns and n rows (where n = number of time steps in the sequencing graph). The 1st column = the time step, 2nd column = operation, 3rd column = Input 1, 4th column = Input 2 and 5th column = Output.

   Repeat for i = 0 to N

   {

2. For time step i (*0<=i<=N, where N is the total number of time steps in the sequencing graph*), check if an operation exists. There could be the following conditions:

   If an **(operation exists in the current time step and there also exists an operation in the next time step)** then assign the respective operation for the current time step i in the column specifying the *operation* of the MST and assign input 1 and input 2 of next operation which would be used in the next time step (i+1) in the columns specifying the *inputs* of the MST. Finally, assign the output of the current operation in the column specifying the *output* of the MST.

   ***Elseif*, (there exists no operation in the current time step and there exists no operation in the next time step),** then assign "-" for the operation in the column specifying the *operation* of the MST, assign "-" for the input 1, assign "-" for the input 2 in the columns specifying the *inputs* of the MST and assign "-" for the output of the current operation in the column specifying the *output* of the MST.

   ***Elseif*, (there exists no operation in the current time step but there exists an operation for the next time step),** then assign "-" for the operation in the column specifying the *operation* of the MST, assign respective inputs for the input 1 and input 2 which would be used in the next time step (i+1), in the columns specifying the *inputs* of the MST. Also, assign "-" for the output of the current operation in the column specifying the *output* of the MST.

   ***Elseif*, (there exists an operation in the current time step but there exists no operation for the next time step),** then assign the respective operation for the current time step i in the column specifying the *operation* of the MST, assign "-" for the input 1, assign "-" for the input 2 in the columns specifying the *inputs* of the MST (since there is no operation in time step i+1, hence no inputs are prepared). Finally, assign the respective output of the current operation in the column specifying the *output* of the MST.

   } **STOP**

Fig. 7.9 Algorithm for converting scheduling to circuit interconnection phase (Note: Reprinted from Anirban Sengupta 2013)

during scheduling. The information in the resource string has resource combination of 1 adder/sub and 2 multipliers. The corresponding scheduling for the DFG in Fig. 7.3 is shown in Fig. 7.8 (Zeng 2010). It is to be noted that assuming one multiplier operation consumes 4 cc and one addition/subtraction consumes 2 cc. The format chosen to represent operation details of each node is: operation_type node# (input 1, input 2). For example, + 12 (10, 11) indicate that the operation type is: addition, node # is: 12, and inputs come from node# 10 and 11, respectively.

7.2.6 Scheduling Phase to Circuit Interconnection Phase

This section proposes an algorithm to directly convert from scheduling phase to circuit interconnection phase represented through multiplexing scheme table (MST). Multiplexing scheme is a procedure for representing each system resource with respective inputs, outputs, operations, and the necessary interconnections (Sengupta et al. 2012). The algorithm is shown in Fig. 7.9. Using this algorithm, the multiplexing scheme Table 7.1 is obtained for adder/subtractor (R_1) from scheduling phase as in Fig. 7.8. Similarly, the multiplexing scheme for other resources can be determined.

Table 7.1 Multiplexing table for adder/sub (A1)

Time slot	Opn	Input 1	Input 2	Output
0	–	–	–	–
1		R_2out	R_3out	
2	(+)	R_1out	R_2out	R_1in
3	(+)	R_1out	Reg 1	R_1in
4	(−)	R_1out	R_2out	R_1in
5	(+)	–	–	RegY

Note: The Table is reconstructed (reused) with the permission of the author in Zeng (2010)

7.2.7 Development of Block Diagram of the Data Path Circuit

Using Table 7.1, the interconnection details are used to develop the block diagram of the data path. The system block diagram consists of data path of the circuit. The data path is accountable for the data flow through the buses. The data path consists of switching components, memory elements, functional resources, and temporary storage elements all working in tandem performing operations concurrently. The block diagram obtained based on the multiplexing scheme information for resources R_1 (Table 7.1), R_2, and R_3 is shown in Fig. 7.10 (Sengupta et al. 2012).

7.2.8 Development of Controller

This major unit prepares the data path units for the incoming data by changing the configuration to perform the next assigned function. For synchronous functioning of all the data elements in the system, the controller has to respond to the requirement exactly at the right instance. Failure to activate or deactivate any element in the data path results in fatal consequences. Controller can be implemented using any hardware description languages.

7.2.9 Register Transfer Level (RTL) Circuit

After the schematic structure of the device has been successfully completed, it is ready for development in any of the available logic synthesis tools. All components in the data path can be described and implemented in VHDL before verification. The schematic structure of the whole device can be drawn and implemented in Xilinx Integrated Software Environment (ISE).

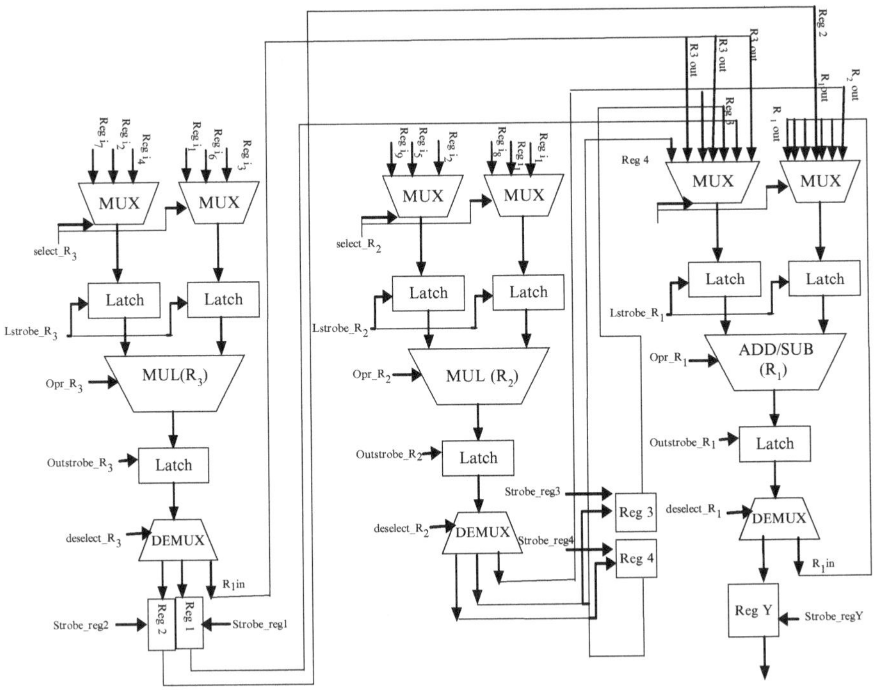

Fig. 7.10 Block diagram of the data path circuit (Note: The figure is reconstructed (reused) with the permission of the author in Zeng 2010)

7.3 Conclusion

This chapter has presented a generic design flow of a data intensive application from an algorithmic level to register transfer level using the evolutionary exploration approach. The chapter provides details of the design steps that can be automated (specially scheduling to data path circuit) and easily integrated with any EA-based exploration approaches available for DFGs.

Acknowledgment I thank IEEE for allowing the reuse of the materials from *Anirban Sengupta "A Methodology for Self Correction Scheme Based Fast Multi Criterion Exploration and Architectural Synthesis of Data Dominated Applications," Proceedings of IEEE International Conference on Advances in Computing, Communications and Informatics (ICACCI-2013), August 2013, pp. 430–436.* for my publication of book chapter.

I would also like to acknowledge the assistance provided by Science and Engineering Research Board (SERB), Department of Science and Technology, Govt. of India for carrying out this work under the sanctioned grant no. SB/FTP/ETA-0474/2012.

I would also like to thank the author, Mr. Zhipeng Zeng, Engineering Officer, Hydro One, Canada for allowing me to reuse materials of his thesis.

References

Dhodhi MK, Hielscher FH, Storer RH, Bhasker J (1995) Datapath synthesis using a problem-space genetic algorithm. IEEE Trans Comput-Aided Des Integr Circuits Syst 14(8):934–944

Gajski D, Dutt ND, Wu A, Lin S (1992) High level synthesis: introduction to chip and system design. Kluwer Academic Publishers, Norwell

Harish Ram DS, Bhuvaneswari MC, Prabhu SS (2012) A novel framework for applying multiobjective GA and PSO based approaches for simultaneous area, delay, and power optimization in high level synthesis of datapaths. VLSI Design, Hindawi, Article ID 273276, 12 pages

Krishnan V, Katkoori S (2006) A genetic algorithm for the design space exploration of datapaths during high-level synthesis. IEEE Trans Evol Comput 10(3):213–229

Mishra VK, Sengupta A (2014) MO-PSE: adaptive multi objective particle swarm optimization based design space exploration in architectural synthesis for application specific processor design. Elsevier J Adv Eng Softw 67:111–124

Papa G, Šilc J (2002) Automatic large-scale integrated circuit synthesis using allocation-based scheduling algorithm. Microprocess Microsyst 26(3):139–147

Sengupta A (2013) A methodology for self correction scheme based fast multi criterion exploration and architectural synthesis of data dominated applications. In: Proceedings of IEEE international conference on advances in computing, communications and informatics (ICACCI-2013), August 2013, Mysore, pp 430–436

Sengupta A, Sedaghat R, Sarkar P (2012) A multi structure genetic algorithm for integrated design space exploration of scheduling and allocation in high level synthesis for DSP kernels. Elsevier J Swarm Evol Comput 7:35–46

Torbey E, Knight J (1998) High-level synthesis of digital circuits using genetic algorithms. In: Proceedings of the international conference on evolutionary computation, Anchorage, AK, pp 224–229

Zeng Z (2010) High level synthesis design flow for multi parametric optimization with hybrid hierarchical design space exploration. Master of Applied Science Thesis, Ryerson University

Chapter 8
Cross-Talk Delay Fault Test Generation

M.C. Bhuvaneswari and S. Jayanthy

Abstract As design trends move toward nanometer technology, new Automatic Test Pattern Generation (ATPG) problems are emerging. During design validation, the effect of cross talk on reliability and performance cannot be ignored. So, new ATPG Techniques have to be developed for testing cross-talk faults that affect the timing behavior of circuits. Further, periodic testing of VLSI circuits can cause generation of excessive heat that can damage the chips under test. Therefore, it is much necessary to reduce the power dissipation during the testing phase also. This chapter deals with two multi-objective genetic-algorithms-based ATPGs, namely, Weighted Sum Genetic Algorithm (WSGA)-based ATPG and Nondominated Sorting Genetic Algorithm (NSGA-II)-based ATPG that generates test pattern set that has high fault coverage and low power consumption. Redundancy is introduced in NSGA-II-based ATPG by modifying the fault-dropping phase and hence very good reductions in transition activity are achieved. Tests are generated for scan versions of ISCAS'89, ISCAS'85, and ITC'99 benchmark circuits. Experimental results demonstrate that NSGA-II-based fault simulator gives higher fault coverage, reduced transitions, and compact test vectors for most of the benchmark circuits when compared with those generated by Weighted Sum Genetic Algorithm (WSGA).

Keywords Crosstalk delay faults • Fault simulator • Optimization algorithms • Test generation • VLSI circuits • Power consumption • Fault coverage • Test vectors • Multi-objective genetic algorithms • NSGA-II

M.C. Bhuvaneswari
Electrical and Electronics Engineering, PSG College of Technology, Coimbatore, India
e-mail: mcb@eee.psgtech.ac.in

S. Jayanthy (✉)
Electronics and Communication Engineering, Sri Ramakrishna College of Engineering, Coimbatore, India
e-mail: jayanthy.s@srec.ac.in

M.C. Bhuvaneswari (ed.), *Application of Evolutionary Algorithms for Multi-objective Optimization in VLSI and Embedded Systems*, DOI 10.1007/978-81-322-1958-3_8,

8.1 Problem Definition

The use of deep submicron process technologies presents several new challenges in the area of manufacturing test. The current deep-submicron technology is based on reduced feature size, increased operating frequencies and clock speeds. The reduction in feature size increases the probability that a manufacturing defect in the Integrated Circuit (IC) will result in a faulty chip. When the feature size is very small, a small defect can result in a faulty transistor or wire. This can cause timing or functional failure of the chip. In any IC manufacturing process physical defects are almost invariably introduced. Hence, some number of ICs is expected to be faulty. Therefore, testing is required to guarantee fault-free electronic systems composed of Very Large Scale Integration (VLSI) devices (Wang et al. 2006).

The major factors that affect the functionality in deep-submicron VLSI designs are process variations, manufacturing defects, and noise. Process variations result in a wide range of device parameters causing difficulty in timing validation and estimation. Delay defects, which are caused by interconnect defects and noise sources such as cross talk, power supply noise, and substrate noise, are difficult to predict. This makes it difficult to estimate the design quality. A lot of design work has been done to tackle the problems arising out of scaling of devices manufactured using deep-submicron technologies. However, efforts to ensure testability of such designs have not been sufficient. More effective fault models are required to model the behavior of faulty deep-submicron chips. Further, there is a significant increase in delay faults in logical paths in ICs. Hence, efficient test techniques are needed to screen out such defects before shipping the products to customers.

Noise is one of the most important factors affecting the deep-submicron VLSI designs. Among the various noise sources, problems related to cross-talk noise between circuits interconnects has become a dominant source of noise in deep-submicron circuits (Gal 1995). High-speed digital circuits heavily use the dynamic logic family. Dynamic circuits with their two-phase clock are more prone to noise effects due to cross talk compared to the static circuits (Heydari and Pedram 2001). As technology approaches submicron era, line and load capacitance dominate circuit behavior. For deep-submicron processes, due to decrease in spacing between conductors, the increase in height-to-width ratio of each conductor, the increase in length of adjoining conductors, and increase in metal layers cross-coupling capacitance dominates circuit behavior. The coupling capacitance becomes a considerable contributor to signal integrity problem.

Coupling causes injection of noise into near lines from active lines. This mechanism of noise coupling is called cross talk. Cross-talk noise may cause undesirable effects including excessive overshoot, undershoot, glitches, additional signal delay, and even a reduction in signal delay (Chen et al. 1997). Two significant problems occur due to cross-talk noise: cross-talk pulse and cross-talk delay. For a pair of nodes with capacitive crosstalk interaction, one of them acts as affecting node or aggressor and the other is the affected or victim node. A cross-talk glitch is a pulse that is induced by coupling effects among interconnects lines.

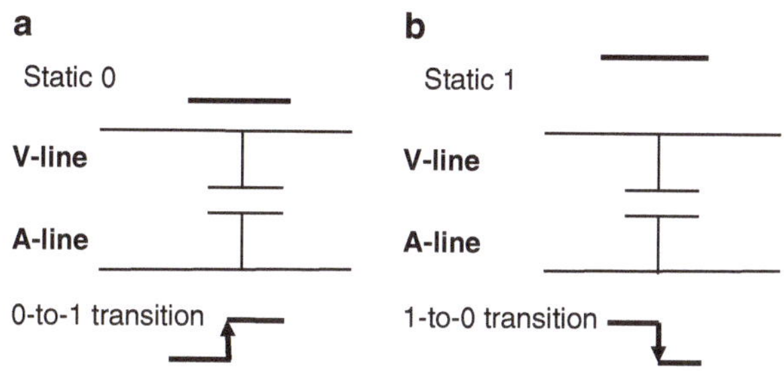

Fig. 8.1 (**a**) Positive glitch. (**b**) Negative glitch

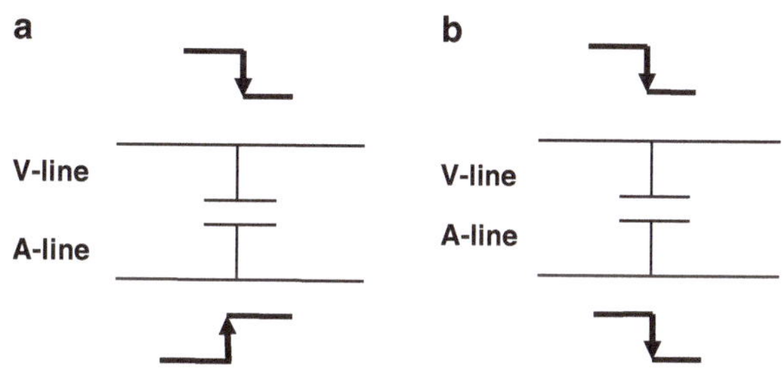

Fig. 8.2 (**a**) Cross-talk slowdown. (**b**) Cross-talk speedup

The amplitude and the width of the pulse depends on aggressors switching time instants, the total value of coupling capacitance, line-to-ground capacitance, and the relative transition times of the aggressors. When a transition signal is applied to aggressor nodes and stable signals are given to the victim node, the victim node will experience coupling noise due to transition in the aggressor nodes. If the strength of the aggressor line is large, then it produces interference in a few other adjacent lines (Pan Zhongliang and Chen Ling 2013).

In Fig. 8.1a, the V-line represents victim line, to which static 0 is applied and a rising transition is applied to one aggressor line (A-line). The aggressor node creates a positive glitch on the victim line, which depending on its amplitude and width will affect the circuit performance (Moll and Rubio 1992). In Fig. 8.1b, static 1 value is given to the victim line and a negative transition, (0–1), is applied to one aggressor line, which causes a negative glitch on the victim line.

Cross-talk delay is a signal delay induced by the cross-talk coupling effects; that is, instantaneous transitions amongst interconnect lines. Cross-talk causes a delay in addition to normal gate delay and interconnects delay. Thus, it is difficult to calculate the correct circuit delay, which may lead to severe signal delay problems (Wang et al. 2006).

In Fig. 8.2a, the victim line has a falling transition and a rising transition is applied at the aggressor and therefore, signal delay increases thus decreasing the circuit speed. On the other hand, as shown in Fig. 8.2b, if the transitions are in the same direction, delay of the signal decreases thus increasing circuit speed. These changes in signal propagation delays can cause faulty behaviors of digital circuits.

When fault analysis tools are designed, these delays have to be taken into consideration. Several physical design (Vittal and Marek-Sadowska 1997; Elgamel et al. 2005; Hunagund and Kalpana 2010) and analysis tools (Agarwal et al. 2006; Kaushik and Sarkar 2008; Palit et al. 2009; Shahin and Pedram 2008) are being developed to aid in designs that minimize cross talk. But, it is impossible to predict in advance the full range of process variations and manufacturing defects which may aggravate capacitance coupling effects. Redesign may also be very expensive in terms of design effort (Ganeshpure and Kundu 2007). Further, cross-talk effects coupled with weak spot defects create situations that must be addressed during testing. Hence, there is a need to develop testing techniques for manufacturing defects that produce cross-talk effects.

Both deterministic and simulation-based algorithms have been proposed for testing cross-talk delay faults. Deterministic test generators are used to produce test vectors by processing a structural model of the circuit. The structural test generation algorithms utilize heuristics to guide the search space. Simulation-based test generators generate random vectors and employ the results provided by the fault simulator to explore the search space. Simulation-based test generation has been used to avoid long execution times of deterministic algorithms and to reduce complexity of test generation. Test generation time can be greatly reduced using efficient fault simulators. Test can be generated for any circuit and any type of fault that can be simulated. Circuit timing can be taken into account.

Genetic Algorithm (GA), which is a simulation-based optimization algorithm, had long been very effective for test generation for both combinational and sequential circuits. The goal of GA in test generation is to find the optimum test vector that detects maximum number of faults. Mazumder and Rudnick (1999) have used GA for simulation-based test generation to minimize execution times and maximize fault coverage.

8.2 Static Timing Analysis

Static timing analysis is one of the techniques available to verify the timing of a digital design. Static timing analysis is used at gate level to derive a list of target faults caused due to cross-talk delay effects. A cross-talk delay fault is caused between an aggressor line denoted as A-line and victim line denoted as V-line. A cross-talk delay fault is induced in V-line during a test when the necessary signal changes occur in A-line. The numbers of cross-talk faults between all possible combinations of A-lines and V-lines in a digital VLSI circuit are very large and impractical to detect for large complex circuits. Therefore, static timing analysis is used to get a reduced set of cross-talk delay faults.

The entire paths in the circuit are analyzed using the tree data structure. The total number of paths in the circuit is calculated using depth first search algorithm, which employs recursive search procedure. From the total number of paths, a set of critical paths whose delay is the largest of the all the paths in the circuit are found.

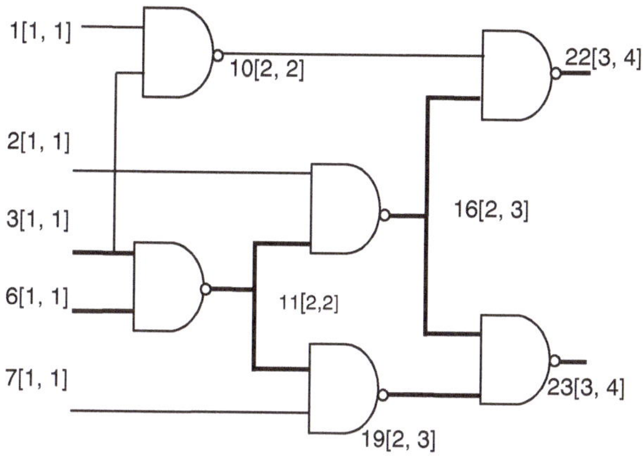

Line number [Earliest transition time, Latest Transition time]

——— Longest paths

Fig. 8.3 Example circuit c17

The victim lines are lines that lie on the critical path. The selection of critical paths, number of victims in the critical path, and the total number of target cross-talk delay faults are found using static timing analysis. The procedure for finding target cross-talk faults consists of two phases. In the first phase, the earliest transition time and latest transition time of each line are calculated. By using this information the longest paths (critical paths) are found. From the critical paths, the set of victim lines are derived. In the second phase, the procedure determines whether the timing window of victim line overlaps with the timing window of the aggressor line. If it overlaps, the pair form a target cross-talk fault (Takahashi et al. 2005).

The method is illustrated by using the ISCAS85 circuit c17 (ACM/SIGDA benchmarks 2007) as shown in Fig. 8.3.

Phase 1. The earliest and latest transition times for each line in circuit c17, as indicated by line number (earliest transition time, latest transition time), are shown in Fig. 8.3. The longest paths for c17 circuit are indicated by bold lines in Fig. 8.3. The victims in the longest path form a set [3, 6, 11, 16, 19, 22, 23].

Phase 2. Select V-line 3 from the victim set. Victim 3 is compared with A-line 1. The timing window of 1(earliest transition time = 1, latest transition time = 1) overlaps with the timing window of 3(latest transition time − 1 = 0, latest transition time + 1 = 2). Hence, (3, 1) form a target cross-talk fault. Considering the pair (3, 22), the timing window of 3(0, 2) does not overlap with timing window of 22 (3, 4). Hence, (3, 22) form a false cross-talk fault and hence should not be included in the target fault list.

The number of target cross-talk faults for benchmark circuit c17 is 42.

8.3 Cross-Talk Delay Fault Simulator

A cross-talk delay fault simulator is used to determine the number of faults detected by the test sequence. A test vector sequence and target fault list derived from static timing analysis are given as inputs to the simulator. The cross-talk delay fault simulator uses 3 logic values 0, 1, x. An event-driven simulation technique (Ulrich 1969) is used for propagating the events. The simulator is capable of processing single-victim/multiple-aggressor faults.

Initially, the simulator reads in a new test sequence. This is followed by good circuit simulation and faults are read from the fault list. All aggressors for a particular victim are read and simultaneously checked for activation. Faults are said to be activated when victim and aggressors have opposite transition. The fault is injected by delaying the victim alone by a time step equal to precalculated delay value. Other events are scheduled as usual as for a good circuit. The simulation is continued until the end of time frame. If delay effect is observed at any primary outputs of faulty circuit, then fault is detected and the victim and activated aggressors pairs are removed from the fault list. The remaining aggressors are considered for the next test pattern. The other faults are read from the fault list and fault simulation is done. The whole process is repeated for all the test vectors in the test set (Jayanthy 2012). Two versions of cross-talk delay fault simulators are developed. In the first version, gates are assumed unit delay and the activated victim is given one unit delay. In the second version, gates are assigned a rise/fall delay and activated victim is given a delay calculated from coupling capacitance and resistance value. The rise/fall delay of gates, coupling capacitance, and resistance value is derived from 45 nm Nangate Generic Open Cell Library (Benchmark site 1) (NANGATE INC 2009).

8.4 Cross-Talk Delay Fault Test Generation Using Multi-objective Genetic Algorithms

To simultaneously optimize two objectives, fault coverage and power consumption, two multi-objective optimization algorithms, Weighted Sum Genetic Algorithm (WSGA) and Nondominated Sorting Genetic Algorithm-II (NSGA-II) (Deb 2001) are used to generate tests for cross-talk delay faults. Redundancy is introduced in the fault-dropping phase of NSGA-II-based Automatic Test Pattern Generation (ATPG).

8.4.1 Chromosome Encoding

The chromosomes are made up of units called genes. Group of chromosomes form a population. Binary encoding scheme is employed. The chromosome representation

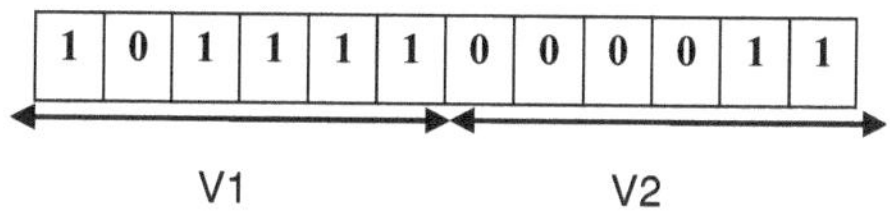

Fig. 8.4 Chromosome encoding

is shown in Fig. 8.4. In cross-talk delay fault testing, a two-vector test pattern, <v1, v2> is required. Hence, each chromosome represents a two-vector pattern. The number of genes in the chromosome is twice the number of primary inputs.

8.4.2 Crossover Operators

Crossover is a GA operator that entails choosing two individuals to swap segments of their string, thereby producing new offspring that are combination of their parents. The one-point, two-point, uniform, and proposed weight-based crossover are described in this section.

8.4.2.1 One-Point Crossover

A single crossover point is randomly selected on both parents. All bits beyond that point in either parent are swapped.

8.4.2.2 Two-Point Crossover

Two-point crossover has two points to be selected on the parent strings. All bits between the two points are swapped, rendering two children.

8.4.2.3 Uniform Crossover

In uniform crossover, the crossover operator decides with some probability which parent will contribute each of the gene values in the offspring chromosomes (Rudnick and Greenstein 1997).

8.4.2.4 Weight-Based Crossover (WCO)

In GA, crossover operators differ from each other primarily in the way they choose the gene from parents to form the offspring. In one-point, two-point, and uniform crossover operators, the selection of the gene is purely random. The WCO operator depends on the weights assigned to the genes for crossover operation. It is a combination of uniform crossover and weighted uniform crossover. If the number of generations is N: uniform crossover is performed for the first N/2 generations and for the next N/2 generations weights are assigned to the genes in the chromosome according to the similarity of the test patterns in the population and weighted uniform crossover is performed (Bhuvaneswari 2001).

8.5 Weighted Sum Genetic Algorithm-Based ATPG

The dual objectives used in WSGA are maximizing fault coverage and minimizing power dissipation. In Complimentary Metal Oxide Semiconductor (CMOS) digital circuits, it has been proved that the parameter that affects the power and energy consumption is the switching activity in the digital circuit. For a given CMOS technology with specified supply voltage of the circuit design, number of switching is the predominant factor that affects the power and energy consumption (Girard 2002). Switching activity of digital circuits is the measure of the number of transitions at primary inputs and gate outputs. Hence, the second objective in WSGA-based ATPG is minimizing the number of transitions.

The weighted sum method is the most widely used classical approach for handling multi-objective optimization problems. It combines a set of objectives into a single objective by premultiplying each objective with a user-supplied weight.

The overall process of test generation using WSGA is shown in Fig. 8.5. An initial population is generated randomly containing individuals of length $2Q$, where Q is the number of primary inputs in each circuit.

The fitness function is:

$$F = (w_1{}^*NF_i) + (w_2{}^*NT_r) \tag{8.1}$$

where NF_i is the number of faults detected and NT_r denotes the number of transitions, the weights $w_1 = 0.6$ and $w_2 = 0.4$. A number of combinations of weights were experimented and these weights were chosen to give optimum results. A cross-talk delay fault simulator is used to find the fitness function.

A pair of individuals is selected using tournament selection, based on the fitness values. Crossover operation is applied for the selected pair of individuals to generate two new individuals with a crossover probability of 1. For each bit value of the individuals generated, mutation operation is applied with a mutation probability of 0.1. Old individuals are removed from the population and replaced with new individuals that have optimal fitness value for the next generation. The algorithm is stopped when the required number of generations is reached.

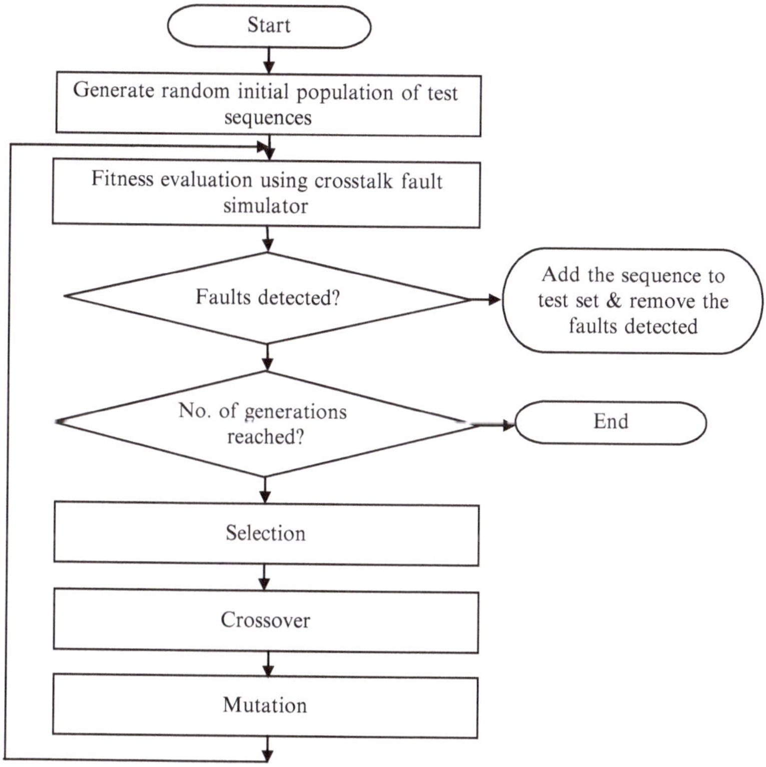

Fig. 8.5 Flowchart of WSGA for cross-talk fault testing

8.6 NSGA-II-Based ATPG

The NSGA-II-based ATPG consists of two phases. In the first phase, the initial population of test sequences is generated based on a pseudo random process. If a sequence detects a fault, that sequence is added to the set, which is given as input to the second phase. In the second phase, child population is created from population obtained in the first phase. To the combined parent and child population, NSGA-II procedure, discussed in Sect. 1.6.2 of Chap. 1 is applied to obtain the new population. The faults that are detected are removed from the fault list and the test sequence is added to the test set. The remaining faults are passed to the next sequence.

The fitness functions used in NSGA-II-based ATPG are given by Eqs. (8.2) and (8.3):

Minimise

$$F_1 = NF_i - FC_i \tag{8.2}$$

$$F_2 = NT_r \tag{8.3}$$

where NF_i is the number of faults that is given to the i-th test sequence, FC_i is the number of faults covered by that test sequence, and NT_r is the number of transitions during the application of the i-th test sequence.

In NSGA-II-based ATPG, a redundancy is introduced in test generation process. The aim in introducing redundancy is to allow NSGA-II to select better test sequences so as to minimize the number of transitions during test pattern generation. Redundancy is introduced by modifying the fault-dropping phase, where a fault is dropped only if it is covered by at least M sequences, where M is the redundancy factor.

8.7 Experimental Results

In this section, the results for test generation using WSGA, NSGA-II, and NSGA-II with redundancy are presented. Tests are generated for most of the ISCAS'85 combinational circuits and several enhanced scan versions of ISCAS'89 sequential circuits. Tables 8.1 and 8.2 give the characteristics of ISCAS'85 combinational

Table 8.1 Characteristics of ISCAS'85 and enhanced scan version of ISCAS'89 benchmark circuits

Circuit name	No. of inputs	No. of outputs	No. of gates	Total no. of paths	No. of longest paths	No. of victims	Total target faults
c17	5	2	6	11	6	7	42
c432	36	7	160	83,926	2,199	103	9,327
c880	60	26	383	8,642	92	70	9,279
c499	41	32	202	9,440	14	207	21,879
c1355	41	32	546	4,173,216	97,372	607	157,387
c1908	33	25	880	729,057	32	93	25,916
c3540	50	22	1,669	28,676,671	95	150	11,252
c5315	178	123	2,307	1,341,305	79	132	60,554
c7552	207	108	3,512	726,493	7	154	138,680
s27	7	4	10	28	6	10	74
s208	19	10	96	145	2	18	743
s208.1	18	9	104	142	1	12	558
s298	17	20	119	231	1	10	537
s344	24	26	160	355	1	21	1,190
s526	24	27	193	410	1	10	891
s386	13	13	159	207	10	49	4,195
s510	25	13	211	369	1	13	1,098
s420.1	34	17	218	474	1	14	1,276
s820	23	24	289	492	11	43	7,738
s953	16	23	395	1,133	2	20	3,036
s1196	32	32	529	3,097	9	55	10,630
s1488	14	25	653	962	1	18	4,305
s1238	32	32	508	3,558	30	45	5,822
s1423	91	79	657	44,726	8	65	15,493
s1494	14	25	647	976	1	18	4,283
s13207.1	700	790	7,951	1,345,319	565	188	241,743

Table 8.2 Characteristics of ITC'99 benchmark circuits

Circuit	No. of PIs	No. of POs	No. of gates	No. of paths	No. of critical paths	No. of victims	Total faults
b01_C	7	7	54	65	3	15	421
b02_C	5	5	32	26	1	7	103
b03_C	34	34	190	791	3	15	2,957
b04_C	77	74	803	84,548	40	65	12,070
b06_C	71	150	65	70	6	15	456
b08_C	30	25	204	2,809	49	43	2,669
b10_C	28	23	223	703	2	21	1,338
b11_C	38	37	801	10,566	2	37	7,845
b14_C	277	299	10,343	57,121,179	92	68	411,412
b20_C	522	512	20,716	289,792,088	193	307	1,625,578
b22_C	767	757	30,686	437,362,328	30	85	812,700

circuits, enhanced scan version of ISCAS'89 sequential circuits, and ITC'99 benchmark circuits. The number of Primary Inputs (PI), number of Primary Outputs (PO), number of gates, number of paths, number of critical paths, number of victims, and total number of target faults corresponding to each circuit are given. The WSGA-based test generator and NSGA-II-based test generator is implemented in INTEL core I7 processor with 2.6 GHz and 4 GB RAM using C language in Linux environment.

WSGA-based ATPG, NSGA-II-based ATPG, and NSGA-II-based ATPG with redundancy is run for the four crossover operators namely one-point crossover, two-point crossover, uniform crossover, and weight-based crossover. During test generation, the number of generations is varied as 4, 8, 12, and 16. For large benchmark circuits, more number of generations is chosen to get better results. Choice of the number of generations and population size is a trade-off between fault coverage, execution time, and number of transitions. A population size of 4, 8, 12, and 16 is used. A crossover probability of 1 and mutation probability of 0.1 is used. The selection scheme used is the binary tournament selection. Redundancy factor is selected carefully because both the test generation time and reduction in number of transitions depend on this. By experimental trial, redundancy factor of 5 is chosen. Unit delay cross-talk delay fault simulator is used to evaluate the fitness function (Jayanthy and Bhuvaneswari 2011).

Experiments are performed using WSGA-based test generation. Tables 8.3 and 8.4, depict the number of faults detected and number of transitions for all the crossover schemes, respectively. The bold number represents the maximum faults detected and minimum number of transitions for each circuit, respectively. The number of faults detected using weight-based crossover is greater than that of the other crossover schemes for 17 of 24 benchmark circuits. The minimum number of transitions for weight-based crossover is obtained for 6 of the 24 benchmark circuits. CPU time is nearly equal for all the crossovers. The results of

Table 8.3 Number of faults detected using WSGA-based ATPG

Circuit	No. of faults detected Uniform	1-Point	2-Point	WCO
c17	**42**	**42**	41	41
c432	7,955	7,946	**7,972**	**7,970**
c880	8,873	8,873	8,876	**8,885**
c499	20,658	20,529	20,669	**20,726**
c1355	127,670	126,220	126,675	**126,701**
c1908	23,998	23,884	23,223	**24,326**
c3540	9,787	9,778	9,870	**9,882**
c7552	130,981	**134,778**	130,034	132,881
s27	66	65	**69**	66
s208	647	630	663	**676**
s208.1	472	528	456	**530**
s298	524	**530**	523	525
s344	1,182	1,185	**1,187**	1,184
s386	4,121	4,112	4,048	**4,136**
s420.1	**1,197**	1,150	1,151	**1,197**
s820	7,636	7,684	7,536	**7,718**
s510	1,073	1,082	1,080	**1,095**
s526	879	**882**	877	**882**
s953	2,876	2,822	**2,968**	2,877
s1196	10,297	10,138	10,162	**10,398**
s1238	4,392	4,514	4,432	**4,614**
s1423	15,374	15,289	15,331	**15,380**
s1488	4,278	4,280	4,296	**4,299**
s13207.1	195,212	195,201	**195,261**	195,245

FAN-based deterministic ATPG (Jayanthy et al. 2013) was taken for comparison. WSGA-based ATPG gave a better fault coverage in reduced CPU time for almost all the benchmark circuits compared with FAN-based ATPG.

Tables 8.5 and 8.6 represent the number of faults detected and number of transitions for each circuit for NSGA-II-based ATPG for all the crossover schemes. It is observed that two-point crossover gives better fault coverage for larger benchmark circuits than other crossover schemes. The number of faults detected using two-point crossover is greater than that of the other crossover schemes for 15 of 24 benchmark circuits. The number of transitions for two-point crossover is minimum for 5 of 24 benchmark circuits.

Experimental results for NSGA-II based ATPG with redundancy is summarized in Tables 8.7 and 8.8. From experimental results, it is observed that two-point crossover gives better fault coverage for 14 of 24 benchmarks. The number of transitions is minimum for two-point crossover for 3 of 24 benchmark circuits.

Table 8.9 shows the number of faults detected and number of transitions for NSGA-II-based ATPG with redundancy for ITC’99 benchmark circuits. The results are observed for two-point crossover.

Table 8.4 Number of transitions using WSGA-based ATPG

Circuit	No. of transitions Uniform	1-Point	2-Point	WCO
c17	363	**369**	386	360
c432	**75,860**	77,464	76,432	78,308
c880	**136,385**	142,222	141,331	137,715
c499	108,632	**107,962**	108,435	108,800
c1355	**293,023**	295,590	297,221	296,609
c1908	484,799	489,909	492,100	**484,691**
c3540	**856,214**	857,121	858,908	859,012
c7552	2,077,961	2,089,081	2,081,300	**2,076,670**
s27	664	663	**636**	663
s208	32,203	**32,095**	32,142	32,207
s208.1	**30,689**	31,086	31,113	31,253
s298	14,129	**13,460**	13,798	14,092
s344	53,016	53,019	**50,302**	52,130
s386	38,698	39,109	39,481	**38,675**
s420.1	58,496	59,368	58,613	**58,483**
s820	75,410	76,109	**75,197**	75,621
s510	58,249	**55,498**	57,245	58,257
s526	53,809	54,294	**53,599**	55,799
s953	83,328	83,328	83,328	83,328
s1196	91,379	90,665	91,172	**90,536**
s1238	58,897	58,208	**58,121**	59,386
s1423	181,431	182,007	**181,113**	181,402
s1488	155,551	160,142	**143,103**	157,125
s13207.1	3,758,980	3,762,214	3,755,541	**3,750,233**

Table 8.10 summarizes the results for NSGA-II-based ATPG with redundancy using unit delay simulator and rise/fall delay simulator. The number of faults detected, number of transitions, number of test vectors, and CPU time are compared. Results are obtained using two-point crossover. The fault coverage decreases for NSGA-II-based ATPG using rise/fall delay fault simulator and hence, number of transitions is also less compared to NSGA-II-based ATPG using unit delay simulator. The CPU time in seconds gives the execution time of the NSGA-II-based test generation. The CPU time for NSGA-II-based ATPG using unit delay simulator is less compared to NSGA-II-based ATPG using rise/fall delay fault simulator.

Table 8.11 shows the percentage increase in fault coverage and percentage reductions in transitions for NSGA-II-based ATPG and NSGA-II-based ATPG with redundancy over WSGA-based ATPG. Results are tabulated for two-point crossover. The bold number represents the maximum number of faults detected and minimum number of transitions obtained for each circuit. It can be observed that with a slight decrease (0.40 %) in the fault coverage, NSGA-II-based ATPG achieves a good reduction in transition activity (17.13 %). When redundancy is introduced in NSGA-II-based ATPG, results observed showed a slight increase in fault coverage (0.58 %) with a high degree of reduction in transition activity

Table 8.5 Number of faults detected using NSGA-II-based ATPG

Circuit	No. of faults detected Uniform	1-Point	2-Point	WCO
c17	**38**	**38**	**39**	**38**
c432	7,951	7,958	7,959	**7,960**
c880	8,654	8,649	8,612	**8,685**
c499	20,689	20,722	**20,724**	20,708
c1355	130,071	130,290	**130,870**	130,492
c1908	22,559	22,610	**22,667**	22,579
c3540	9,884	9,801	**9,912**	9,901
c7552	133,194	132,998	**133,318**	130,299
s27	70	70	69	70
s208	608	650	**709**	**670**
s208.1	500	493	492	**509**
s298	**525**	522	**526**	524
s344	1,182	**1,185**	1,181	1,182
s386	**4,160**	4,142	4,116	4,101
s420.1	**1,076**	1,001	1,001	1,016
s820	7,637	7,580	**7,693**	7,650
s510	1,073	1,070	**1,081**	**1,081**
s526	881	881	881	881
s953	2,824	2,874	**2,998**	2,868
s1196	9,479	9,518	**9,607**	9,606
s1238	4,134	4,162	**4,171**	4,167
s1423	15,411	15,346	**15,416**	15,373
s1488	3,980	4,084	**4,201**	4,060
s13207.1	196,839	196,630	**196,910**	196,872

Table 8.6 Number of transitions using NSGA-II-based ATPG

Circuit	No. of transitions Uniform	1-Point	2-Point	WCO
c17	252	243	**244**	253
c4932	**64,590**	67,806	67,556	68,675
c880	**125,087**	126,624	126,448	127,934
c499	**94,511**	95,561	95,094	95,533
c1355	**264,103**	264,745	265,652	265,409
c1908	434,799	435,690	**433,188**	435,299
c3540	**776,723**	776,970	779,980	777,331
c7552	1,837,961	**1,836,390**	1,868,900	1,839,088
s27	419	399	398	**393**
s208	**25,137**	25,839	26,052	26,013
s208.1	**23,572**	23,761	23,622	23,797
s298	**12,355**	12,535	12,647	12,749
s344	43,731	43,850	44,873	**43,047**
s386	**33,207**	34,326	33,591	34,273
s420.1	48,407	48,487	**48,328**	50,023
s820	**64,435**	65,900	79,153	66,641
s510	**47,495**	48,813	49,979	48,982
s526	49,332	48,727	**47,883**	48,648
s953	69,440	69,440	**69,435**	69,440
s1196	68,824	**67,426**	69,142	68,648
s1238	**41,951**	42,071	42,381	42,033
s1423	**145,967**	146,251	148,472	148,113
s1488	114,879	117,420	117,028	**114,537**
s13207.1	3,236,970	3,237,611	**3,236,402**	3,236,581

Table 8.7 Number of faults detected using NSGA-II-based ATPG with redundancy

Circuit	No. of faults detected			
	Uniform	1-Point	2-Point	WCO
c17	42	41	41	41
c432	8,585	8,585	**8,589**	8,588
c880	8,650	8,619	8,640	**8,718**
c499	20,529	20,585	**20,590**	20,570
c1355	128,539	128,401	**128,609**	128,522
c1908	24,081	**24,440**	24,392	24,221
c3540	9,920	9,923	**9,931**	9,926
c7552	133,096	133,220	**133,431**	133,303
s27	70	**72**	**72**	70
s208	**713**	663	680	652
s208.1	523	**558**	515	522
s298	525	**528**	518	526
s344	1,186	**1,189**	1,187	1,187
s386	4,152	4,153	**4,161**	4,159
s420.1	1,112	1,042	**1,080**	1,077
s820	7,711	**7,717**	7,700	7,700
s510	**1,092**	**1,092**	1,090	1,087
s526	879	883	**885**	884
s953	2,824	2,976	**2,990**	2,835
s1196	9,754	9,676	**9,688**	9,751
s1238	**4,585**	4,174	4,094	4,110
s1423	15,380	15,323	**15,414**	15,390
s1488	4,238	4,232	**4,261**	4,231
s13207.1	196,885	197,001	**197,389**	196,302

(27.87 %). Redundancy is introduced by modifying the fault-dropping phase, where a fault is dropped only if it is covered by five sequences. The number of test vectors that detected faults and CPU time is slightly greater when redundancy is introduced compared with NSGA-II-based ATPG without redundancy.

Table 8.12 shows percentage increase in fault coverage and percentage reductions in transitions for NSGA-II-based ATPG and NSGA-II-based ATPG with redundancy over WSGA for larger benchmark circuits. The bold number represents the maximum number of faults detected and minimum number of transitions obtained for each circuit. It is observed that with a slight increase in the fault coverage (0.78 %), NSGA-II-based ATPG achieves a good reduction in transition activity (12.46 %). When redundancy is introduced in NSGA-II-based ATPG, results observed showed an increase in fault coverage (1.88 %) with a high degree of reduction in transition activity (26.51 %). The experimental results indicate that NSGA-II-based ATPG with redundancy gives better fault coverage with minimum number of transitions for all the benchmark circuits.

Figure 8.6 shows the comparison between the execution time of NSGA-II-based ATPG and WSGA-based ATPG. The execution time of WSGA-based ATPG is higher for all the benchmark circuits compared to NSGA-II-based ATPG.

For larger benchmark circuits like c7552, the time for finding the target fault list and execution time of the algorithms is high.

Table 8.8 Number of transitions using NSGA-II-based ATPG with redundancy

Circuit	No. of transitions			
	Uniform	1-Point	2-Point	WCO
c17	217	178	206	**176**
c432	**58,644**	58,666	60,387	62,030
c880	**108,900**	112,951	112,489	112,848
c499	**87,282**	89,094	88,945	87,890
c1355	**236,176**	236,503	237,242	236,612
c1908	353,844	354,899	**352,765**	354,009
c3540	**566,111**	566,801	567,120	566,248
c7552	1,641,854	1,642,843	1,642,982	**1,641,795**
s27	287	285	325	**279**
s208	**22,792**	23,009	23,183	23,379
s208.1	20,912	**20,655**	21,025	21,046
s298	**9,408**	9,471	9,420	9,650
s344	**38,435**	39,516	39,738	41,308
s386	**29,288**	29,982	29,476	30,609
s420.1	**44,223**	44,412	44,463	44,602
s820	58,896	58,941	**58,626**	59,549
s510	**43,431**	44,353	46,216	45,702
s526	**43,480**	44,063	43,990	43,634
s953	62,496	62,496	62,496	62,496
s1196	62,829	62,311	62,065	**60,875**
s1238	36,539	36,023	**35,663**	36,337
s1423	134,068	**126,614**	132,165	129,293
s1488	95,067	**94,541**	98,884	98,089
s13207.1	2,668,874	**2,650,021**	2,655,410	2,641,387

Table 8.9 Experimental results for NSGA-II-based ATPG with redundancy for ITC'99 benchmark circuits

Circuit	No. of faults detected	No. of transitions	CPU time (secs)
b01_C	313	599	0.067
b02_C	87	347	0.017
b03_C	2,790	5,533	1.58
b04_C	10,339	28,109	302.7
b06_C	454	17,174	0.233
b08_C	2,452	14,213	1.95
b10_C	1,265	16,245	0.917
b11_C	3,670	61,125	47.9
b14_C	284,941	2,931,817	11095.3
b20_C	1,095,709	11,218,268	43112.5
b22_C	372,657	11,239,573	38231.7

Table 8.10 Results for NSGA-II-based ATPG with redundancy using unit delay and rise/fall delay simulator

	Unit delay				Rise/fall delay			
Circuit	No. of faults detected	No. of transitions	No. of test vectors	CPU time (sec)	No. of faults detected	No. of transitions	No. of test vectors	CPU time (sec)
c17	41	206	30	0.017	32	236	32	0.33
c880	8,640	112,489	384	13.2	6,717	97,688	384	30.85
c499	20,590	88,945	544	10.3	8,283	41,714	512	21.16
c1355	128,609	237,242	544	410.21	74,538	91,220	512	456.02
c1908	24,392	352,765	544	330.5	3,696	156,636	512	360.9
c3540	9,931	567,120	544	760.2	3,723	389,300	512	800.54
c7552	133,431	1,642,982	544	1089.5	57,641	1,166,517	512	1327.2
s27	72	325	30	0.017	63	279	32	0.33
s208	680	23,183	256	0.417	624	22,698	384	1.56
s208.1	515	21,025	320	0.43	433	20,188	320	1.53
s298	518	9,420	128	0.27	485	8,529	128	0.583
s344	1,187	39,738	320	1.18	720	15,163	320	1.41
s386	4,161	29,476	320	1.9	3,853	25,404	320	4.7
s420.1	1,080	44,463	320	1.3	1,043	42,404	320	4.68
s820	7,700	58,626	320	5.35	7,377	55,725	320	21.6
s510	1,090	46,216	320	0.86	1,039	39,704	320	1.4
s526	885	43,990	320	0.88	763	40,723	320	4.5
s953	2,990	62,496	160	3.3	1,562	30,187	160	5.067
s1196	9,688	62,065	192	7.4	7,535	32,097	192	13.5
s1238	4,094	35,663	128	8.3	2,168	22,880	128	11.15
s1423	15,414	132,165	320	33.4	12,901	98,546	320	64.1
s1488	4,261	98,884	256	4.2	3,360	68,637	256	9.16
s13207.1	197,389	2,655,410	544	3316.1	91,808	2,289,146	512	3568.2

8.8 Summary

In this chapter, multi-objective evolutionary algorithms, Weighted Sum Genetic Algorithm and Nondominated Sorting Genetic Algorithm-II, are used for testing cross-talk delay faults in VLSI circuits. Redundancy is introduced in the fault-dropping phase of NSGA-II-based ATPG and results are analyzed. Experimental results for ISCAS'85 combinational and ISCAS'89 enhanced scan version of sequential circuits and ITC99 benchmark circuits are presented. NSGA-II-based ATPG with redundancy increases the fault coverage with reduced switching activity. The experimental results demonstrate the effectiveness of the proposed approach. The method achieves 81–97 % fault coverage on combinational circuits and 70–99 % fault coverage on enhanced scan version of sequential circuits with an average of 27 % reduction in the number of transitions on combinational and enhanced scan version of sequential circuits. Compared to WSGA-based ATPG, NSGA-II-based ATPG with redundancy reduces the number of transitions with a slight increase in fault coverage for most of the benchmark circuits.

Table 8.11 Comparison of fault coverage and number of transitions

	WSGA-based ATPG		NSGA-II-based ATPG				NSGA-II-based ATPG with redundancy			
Circuit name	% Fault coverage	No. of transitions	% Fault coverage	No. of transitions	Decrease in % fault coverage over WSGA	Reduction in no. of transitions over WSGA	% Fault coverage	No. of transitions	Increase in % fault coverage over WSGA	Reduction in no. of transitions over WSGA
c17	**97.62**	386	92.86	244	4.76	36.78	97.61	**206**	−0.01	**46.63**
c432	85.47	76,432	85.33	67,556	0.14	11.61	**92.08**	**60,387**	6.51	**20.99**
c880	**95.66**	141,331	92.81	126,448	2.85	10.53	93.11	**112,489**	−2.55	**20.4**
c449	94.47	108,435	**94.72**	95,094	−0.25	12.3	94.1	**88,945**	−0.37	**17.97**
s27	91.19	636	**93.24**	398	−2.05	37.42	**97.29**	**325**	6.1	**48.89**
s208	90.98	32,142	**95.42**	26,052	−4.44	18.94	**91.52**	**23,183**	0.54	**27.87**
s208.1	81.72	31,113	**88.17**	23,622	−6.45	24.07	**92.29**	**21,025**	10.57	**32.42**
s298	**97.39**	13,798	**97.95**	12,647	−0.46	8.34	96.46	**9,420**	−0.93	**31.72**
s344	99.74	50,302	99.24	44,873	0.5	10.79	**99.74**	**39,738**	0	**21**
s386	96.49	39,481	**98.12**	33,591	−1.63	14.91	**99.18**	**29,476**	2.69	**25.34**
s420.1	**90.2**	58,613	78.49	48,328	11.71	17.54	84.63	**44,463**	−5.57	**24.14**
s820	97.39	75,197	**99.42**	79,153	−2.03	−5.26	**99.5**	**58,626**	2.11	**22.03**
s510	98.36	57,245	**98.45**	49,979	−0.09	12.69	**99.27**	**46,216**	0.91	**19.26**
s526	98.43	53,599	**98.74**	47,883	−0.31	10.66	**99.32**	**43,990**	0.89	**17.92**
s953	97.76	83,328	**98.78**	69,435	−1.02	16.67	**98.48**	**62,496**	0.72	**25**
s1196	**95.5**	91,172	90.74	69,142	4.76	24.16	91.13	**62,065**	−4.37	**31.92**
s1238	**76.12**	58,121	71.64	42,381	4.48	27.08	70.31	**35,663**	−5.81	**38.64**
s1423	98.95	181,113	**99.5**	148,472	−0.55	18.02	**99.49**	**132,165**	0.54	**27.02**
s1488	**99.79**	143,103	97.58	117,028	2.21	18.22	98.98	**98,884**	−0.81	**30.9**
Average					**0.4**	**17.13**			**0.58**	**27.89**

Table 8.12 Comparison of fault coverage and number of transitions for larger benchmarks

Circuit	WSGA-based ATPG		NSGA-II-based ATPG				NSGA-II-based ATPG with redundancy			
	Fault coverage (%)	No. of transitions	Fault coverage (%)	No. of transitions	Increase in fault coverage over WSGA (%)	Decrease in transitions over WSGA (%)	Fault coverage (%)	No. of transitions	Increase in fault coverage over WSGA (%)	Decrease in transitions over WSGA (%)
c1355	80.48	297,221	**83.15**	264,103	2.67	11.14	81.71	**237,242**	1.23	20.17
c1908	89.60	492,100	87.46	434,799	−2.19	11.64	**94.11**	**353,844**	4.31	28.09
c3540	87.71	358,908	88.09	776,723	0.38	14.02	**88.25**	**567,120**	0.53	33.97
c7552	93.76	2,081,300	96.13	1,837,961	2.37	11.69	**96.21**	**1,642,982**	2.45	21.05
s13207.1	80.77	3,755,541	81.45	3,236,402	0.68	13.82	**81.65**	**2,655,410**	0.88	29.29
Average					**0.78**	**12.46**			**1.88**	**26.51**

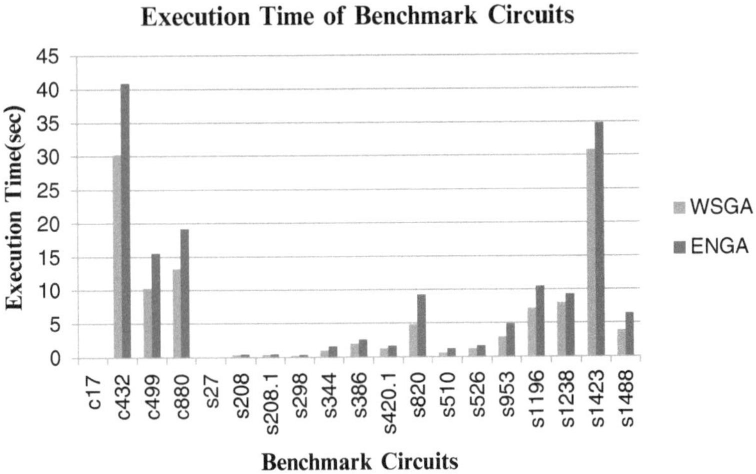

Fig. 8.6 Comparison of execution time

References

ACM/SIGDAbenchmarks. http://www.cbl.ncsu.edu/benchmarks/Benchmarks-upto-1996.html. Accessed Dec 2007

Agarwal K, Sylvester D, Blaauw D (2006) Modeling and analysis of crosstalk noise in coupled RLC interconnects. IEEE Trans Comput Aided Des Integr Circuits Syst 25(5):892–901

Bhuvaneswari MC (2001) Development of new algorithms for test generation and simulation of stuck-at-faults in logic circuits. Dissertation, Bharathiar University, Coimbatore

Chen WY, Gupta SK, Breuer MA (1997) Analytic models for crosstalk delay and pulse analysis for non ideal inputs. In: Proceedings of the international test conference, Washington, DC, USA, pp 809–818

Deb K (2001) Multi-objective optimization using evolutionary algorithms. Wiley, Chichester

Elgamel MA, Kumar A, Bayoumi MA (2005) Efficient shield insertion for inductive noise reduction in nanometer technologies. IEEE Trans Very Large Scale Integr (VLSI) Syst 13(3): 401–405

Gal L (1995) On-chip cross talk – the new signal integrity challenge. In: Proceedings of IEEE custom integrated circuit conference, Santa Clara, California, 1–4 May 1995

Ganeshpure KP, Kundu S (2007) On ATPG for multiple aggressor crosstalk faults in presence of gate delays. In: Proceedings of IEEE international test conference, Santa Clara, California, USA, pp 21–26

Girard P (2002) Survey of low-power testing of VLSI circuits. IEEE Des Test Comput 19(3):82–92

Heydari P, Pedram M (2001) Analysis and reduction of capacitive coupling noise in high-speed VLSI circuits. In: Proceedings of international conference on computer design, Austin, TX, 23–26 September 2001

Hunagund PV, Kalpana AB (2010) Analytical noise modeling for shielding to reduce crosstalk noise in on-chip interconnects. Int J Comput Sci Netw Secur 10(11):19–23

Jayanthy S (2012) Development of algorithms for test generation and simulation of crosstalk delay faults in VLSI circuits. Dissertation, Anna University

Jayanthy S, Bhuvaneswari MC (2011) An efficient multi-objective genetic algorithm for low power testing of crosstalk delay faults in VLSI circuits. Adv Model Anal 54(2):28–48

Jayanthy S, Bhuvaneswari MC, Prabhu M (2013) Simulation based ATPG for low power testing of crosstalk delay faults in asynchronous circuits. Int J Comput Appl Technol 48(3):241–252

Kaushik BK, Sarkar S (2008) Crosstalk analysis for a CMOS-gate-driven coupled interconnects. IEEE Trans Comput Aided Des Integr Circuits Syst 27(6):1150–1154

Mazumder P, Rudnick EM (1999) Genetic algorithms for VLSI design, layout &test automation. Prentice Hall PTR, New York

Moll F, Rubio A (1992) Spurious signals in digital CMOS VLSI circuits: a propagation analysis. IEEE Trans Circuits Syst II Analog Digital Signal Process 39(10):749–752

NANGATE INC (2009) Nangate 45nm open cell library. http://www.nangate.com/index.php?option=com_content&task=view&id=137&Itemid=137. Accessed Nov 2009

Palit AK, Hasan S, Anheier W (2009) Decoupled victim model for the analysis of crosstalk noise between on-chip coupled interconnects. In: Proceedings of 11th electronics packaging technology conference, Singapore, 9–11

Pan Zhongliang, Chen Ling (2013) An approach to generate tests for multiple victim lines of crosstalk faults in integrated circuits. J Theor Appl Inf Technol 50(2):455–461

Rudnick EM, Greenstein GS (1997) A genetic algorithm framework for test generation. IEEE Trans Comput Aided Des Integr Circuits Syst 16(9):1034–1044

Shahin N, Pedram M (2008) Crosstalk-affected delay analysis in nanometer technologies. Int J Electron 95(9):903–937

Takahashi H, Keller KJ, Le KT et al (2005) A method for reducing the target fault list of crosstalk faults in synchronous sequential circuits. IEEE Trans Comput Aided Des Integr Circuits Syst 24(2):252–263

Ulrich EG (1969) Exclusive simulation of activity in digital networks. Commun ACM 12(2):102–110

Vittal A, Marek-Sadowska M (1997) Crosstalk reduction for VLSI. IEEE Trans Comput Aided Des Integr Circuits Syst 16:290–298

Wang L-T, Wu C-W, Wen X (2006) VLSI test principles and architectures: design for testability. Morgan Kaufmann, San Francisco

Chapter 9
Scheduling in Heterogeneous Distributed Systems

M.C. Bhuvaneswari and G. Subashini

Abstract Parallel and distributed systems play an important part in the improvement of high-performance computing. In analyzing the performance of such Heterogeneous Distributed Systems (HDS), scheduling a set of tasks to the available set of resources for execution is highly important. Task-scheduling being an NP-complete problem, use of meta-heuristics is more appropriate in obtaining optimal solutions. The problem of scheduling an application comprising a set of independent tasks on a fully connected HDS is considered here. The scheduling algorithms for this problem focus on minimizing two objectives, the make-span and the flow-time. These objectives conflict with one another, which requires multi-objective problem formulation. Multi-objective Genetic Algorithms (MOGAs) and Multi-objective Particle Swarm Optimization (MOPSO) algorithms are applied to the problem. Weighted Sum Genetic Algorithm (WSGA) based on weighted sum approach, and Nondominated Sorting Genetic Algorithm (NSGA2), a modified version of NSGA2 with controlled elitism (NSGA2-CE) and a hybrid version of NSGA2 combined with Pareto hill climbing (PHC), the NSGA2-PHC are applied to the problem. A weighted sum particle swarm optimization (WSPSO), a Nondominated Sorting Particle Swarm Optimization (NSPSO), an Adaptive Nondominated Sorting Particle Swarm Optimization (ANSPSO), and a hybrid version of ANSPSO with PHC, the NSPSO-PHC are the MOPSO methods applied to task scheduling. The different MOGA and MOPSO methods are compared among themselves to determine the algorithms that generate the efficient schedule. The best version of MOGA is compared with the best MOPSO method to find that MOGA produces better schedules for benchmark test instances simulated.

M.C. Bhuvaneswari
Electrical and Electronics Engineering, PSG College of Technology, Coimbatore, India
e-mail: mcb@eee.psgtech.ac.in

G. Subashini (✉)
Information Technology, PSG College of Technology, Coimbatore, India
e-mail: suba@ity.psgtech.ac.in

M.C. Bhuvaneswari (ed.), *Application of Evolutionary Algorithms for Multi-objective Optimization in VLSI and Embedded Systems*, DOI 10.1007/978-81-322-1958-3_9,

Keywords Heterogeneous computing • Task-scheduling • Multi-objective optimization • Pareto-optimal • NSGA-II • NSPSO • Controlled elitism • Adaptive NSPSO • Pareto hill climbing

9.1 Introduction

The availability of powerful computers and high-speed networks as low-cost commodity components are changing the way computers are used. These advances in technology have led to the emergence of Heterogeneous Computing (HC) environments that utilize a set of interconnected high-performance computers physically distributed to solve large-scale applications in science and engineering (Freund and Siegel 1993; Eshaghian 1996; Foster and Kesselman 2003). Some examples of HC systems include computer clusters, network of workstations, and computational grids.

To realize effective and efficient utilization of the underlying resources and exploiting parallelism inherent in the tasks to be executed, scheduling algorithms are fundamentally important. Scheduling plays an important role in increasing the system performance in a wide range of application domains such as logistics, telecommunications, production processes, mobile devices, parallel computers, and grid computing. The scheduling problem that deals with assignment of tasks to the available resources in general is NP-complete (Garey and Johnson 1979; Coffman 1976) and difficult to solve even in its simplest form. Problem complexity increases when the application is executed on a Heterogeneous Distributed System (HDS) as both the computational nodes and the underlying networks connecting them are heterogeneous. Heterogeneity results in different capabilities for task execution, data accessing, and transmission.

The scheduling problem exists in two forms: static and dynamic. Static scheduling, which is usually done at compile time, requires information regarding the application and computing resources such as execution time, communication cost, data dependency, and synchronization requirement in advance (James 1999). Dynamic scheduling is done at run time and uses more realistic assumptions (Page et al. 2010). But, it suffers from the drawback of run-time overhead due to inappropriate allocation. The wide applicability of static scheduling in different types of analyses and environments make it an important area for ongoing research. The focus of this research is on static scheduling.

As the scheduling problem deals with the selection of the best schedule of all valid schedules, an objective function is required to compare different possible schedules. Many task-scheduling problems discussed in the literature use simple scalar evaluation objectives. But in real situations, multiple factors such as time, monetary cost, reliability, fault tolerance, and security are to be considered while making a scheduling decision to achieve maximum efficiency of the system (Subashini and Bhuvaneswari 2012; Liu et al. 2010). However, with the increase in the number of objectives to be considered, it becomes infeasible to develop a simple heuristic to handle the task-scheduling optimization problem. Therefore, it is necessary to investigate meta-heuristic approaches (Krömer et al. 2010) that are capable of being applied to complex domains. Multi-objective Evolutionary Algorithms (MOEAs) have become the research focus due to their recent popularity in solving multi-objective optimization problems in many application domains, such as control systems (Fleming and Purshouse 2002) and network routing (Roy and Das 2002). These algorithms are capable of simultaneously optimizing multiple objectives without combining them into a single scalar-objective function. Task-scheduling problem for HDS is thus formulated as a multi-objective optimization problem (Deb 2001) as it involves simultaneous optimization of multiple objectives that often are competing or conflicting.

This chapter focuses on developing algorithms for scheduling applications that comprise a set of independent tasks on a Heterogeneous Distributed System (HDS). The algorithms develop schedules considering minimization of two objectives, namely, make-span and flow-time simultaneously. The mathematical formulation of the multi-objective problem along with the objectives to be optimized is described. Application of multi-objective genetic algorithms (MOGAs), WSGA, NSGA-II (Deb 2001), NSGA-II-CE, NSGA-II-PHC, to the task-scheduling problem is discussed. Another evolutionary technique PSO (Lei Zhang et al. 2008) and its multi-objective variants WSPSO, NSPSO, ANSPSO, and NSPSO-PHC are applied to task scheduling. The simulation details of the data set used in testing is explained. The experimental analysis is done and the result of Pareto-based methods is compared with weight-based method. The effectiveness of the Pareto-based algorithms is compared by evaluating them against three performance metrics: Ratio of Nondominated Individuals (*RNI*), Maximum Spread (S_{max}), and Coverage of Two Sets (*C*).

9.2 Task-Scheduling Model

A heterogeneous distributed environment is composed of computational units that can be a single personal computer, a group of workstations, or a supercomputer. The problem is formally defined as follows. Let *P* be the set of *n* processors in the heterogeneous computing system and *T* be the set of *m* tasks to be assigned to the processors. The tasks are assumed to be independent of each other and there is no precedence relation among the tasks. Each task can be assigned only to a single processor and this is executed exclusively. It is also assumed that preemption of tasks is not allowed. As the scheduling is performed statically, an estimation of the computational load of each task, the computing capacity of each processor, and an estimation of the prior load of each processor are required (Braun et al. 2001).

Having the computing capacity of the processor and the workload of the tasks, an Expected Time to Compute (ETC) matrix can be built (Ali et al. 2000). An ETC matrix is an $m \times n$ matrix, where each element ETC[m][n] indicates the Expected Time to Compute task m in processor n. One row of the ETC matrix specifies the

Table 9.1 Example of an ETC matrix

Task	Processor 1	Processor 2
Task 1	3 tu	2 tu
Task 2	2 tu	4 tu
Task 3	7 tu	5 tu
Task 4	6 tu	7 tu

estimated execution time in same time units (tu) for a given task on each processor. Similarly, one column of the ETC matrix consists of the estimated execution time on a given processor for each task. A simple example of an ETC matrix for a system containing 4 tasks and 2 processors is given in Table 9.1.

An instance of the problem specifies the number of independent tasks that must be scheduled, number of heterogeneous processors in HDS, and the Expected Time to Compute, ETC, where the position ETC[m][n] indicates the expected execution time of task m in processor n. This matrix is explicitly provided.

9.3 Scheduling Objectives

To rate the quality of the schedules, an evaluation function that optimizes two objectives make-span and the flow-time is used (Carretero et al. 2007). Make-span is an indicator of the general throughput of the system. Small values of make-span mean that the scheduler is providing good and efficient planning of tasks to resources. On the other hand, flow-time refers to the response time to the user requests for task execution. Minimizing the value of flow-time therefore means reducing the average response time of the system. It is essential to maximize the throughput of the system and at the same time offer a quality of service acceptable to the users. To minimize flow-time, scheduling of Shortest Job on the Fastest Resource (SJFR) is required and make-span minimization includes scheduling the Longest Job on the Fastest Resource (LJFR). Thus, minimization of make-span results in maximization of flow-time. The relationship between the objectives being contradictory makes the problem multi-objective.

The objective function for an optimal schedule is defined as follows:

$$\text{Minimize } F = (MS,\ FT)$$

where the make-span *MS* or finish time is the maximum value of completion time of all tasks and *FT*, the flow-time, is the sum of completion times of all the tasks given by Eqs. 9.1 and 9.2:

$$MS = \left\{ \max_{t \in \text{Tasks}} \text{Finish}_t \right\} \tag{9.1}$$

$$FT = \left\{ \sum_{t \in \text{Tasks}} \text{Finish}_t \right\} \tag{9.2}$$

where Tasks stands for the set of all tasks and Finish_t represents the time by which task t finishes.

The make-span and flow-time values are in incomparable ranges as flow-time has an order of magnitude higher than the make-span. This difference increases with increase in the number of tasks and the processors. So, mean flow-time (MFT), as given by Eq. 9.3, is used to evaluate the flow-time:

$$\text{MFT} = FT/n \tag{9.3}$$

where FT is the flow-time given by Eq. 9.2 and n represents the number of processors in the heterogeneous system.

9.4 Schedule Representation

Each chromosome in a GA is represented as a vector of length equal to the number of tasks. The values specified in this vector are in the range (1, number of processors) and the value corresponding to each position n in the vector represents the processor to which the task n is allocated. A processor number can appear more than once in the schedule if both the tasks are assigned to the same processor. Considering the case of 7 tasks and 3 processors, a schedule representation and the task assignment is as shown in Fig. 9.1.

9.4.1 Generating the Initial Population

The initial population for the multi-objective GA and PSO methods applied for scheduling independent tasks is generated randomly to contain predefined number

a

1	3	2	1	2	1	3
Task 1	Task 2	Task 3	Task 4	Task 5	Task 6	Task 7

b

Processor 1	Task 1	Task 4	Task 6
Processor 2	Task 3	Task 5	
Processor 3	Task 2	Task 7	

Fig. 9.1 GA chromosome (**a**) Schedule representation (**b**) Task assignment

of individuals or schedules. Besides random initialization, one schedule is generated using LJFR-SJFR (Abraham et al. 2008) heuristic to seed the initial population. This heuristic tries to optimize alternatively both the make-span and the flow-time thereby enabling greater convergence. LJFR assigns the largest task to the fastest resource, thereby reducing make-span, whereas SJFR assigns the smallest task to the fastest resource to minimize flow-time. Both techniques are alternatively applied till the set of tasks are completely scheduled.

9.4.2 *Selection*

The multi-objective genetic algorithms make use of a crowded tournament selection operator used by NSGA-II discussed in Sect. 1.6.2 of Chap. 1.

9.4.3 *Mutation*

The multi-objective genetic algorithms perform mutation to assist the search escape from local optima. The mutation operator proposed for the problem introduces a heuristic into the mutation process to improve the performance of GA (Xhafa et al. 2007). The schedule to be mutated is selected and a random position is chosen in the selected schedule. The task in the selected position is reassigned to a processor, which results in minimal completion time of the task (Subashini 2013). The same mutation operators are used in the Pareto hill-climbing stage of the hybrid NSGA-II-PHC and NSPSO-PHC algorithms.

9.5 Performance Measures

To determine the performance of an MOEA, a variety of quantitative measures are suggested to express the quality of solutions (Zitzler et al. 2000). Three such important goals to be achieved by MOEA is to

1. Maximize the number of elements of the Pareto-optimal set found.
2. Maximize the spread of solutions found so that a distribution of vectors as smooth and uniform as possible can be obtained.
3. Minimize the distance of the Pareto front produced by algorithm with respect to the global Pareto front (assuming its location is known).

Few metrics such as Generational Distance (GD), Spacing, and Maximum Pareto-optimal Front Error (MFE) are not used here as they require the knowledge of true Pareto-optimal front which is not known. The algorithms are evaluated for the metrics, Ratio of Nondominated Individuals (*RNI*), Maximum Spread (S_{max}), and Coverage of Two sets (*C*). They are described as follows.

9.5.1 Ratio of Nondominated Individuals

The main goal of any MOEA is to obtain the maximum number of possible candidate solutions known as the Pareto-optimal set front from a given population. This performance measure is denoted here as the Ratio of Nondominated Individuals (RNI) for a given population and is mathematically formulated as in Eq. 9.4:

$$\mathrm{RNI}(P) = \frac{\mathrm{nondom}_{\mathrm{indiv}}}{N} \tag{9.4}$$

where $\mathrm{nondom}_{\mathrm{indiv}}$ is the number of nondominated individuals in population P and N is the size of population P. The value $\mathrm{RNI} = 1$ means that all the individuals in the population are nondominated, while $\mathrm{RNI} = 0$ represents the situation where none of the individuals in the population are nondominated.

9.5.2 Maximum Spread

The Maximum Spread S_{max} is the index for measuring the extent of the search space. It calculates the Euclidean distance between the maximum and the minimum of each function. The Maximum Spread may be written as in Eq. 9.5:

$$S_{\max} = \sqrt{\sum_{i=1}^{n} \left| f_i^{\max} - f_i^{\min} \right|} \tag{9.5}$$

S_{max}: the maximum Euclidean distance
$f_i^{\max(\min)}$: max (min) value of the i-th objective function

Greater the value of Maximum Spread indicates coverage of large search space, thereby achieving greater diversity among solutions.

9.5.3 Coverage of Two Sets

This measure can be used to show that the outcomes of one algorithm dominate the outcomes of another algorithm. The *Coverage of Two Sets (C)* is a measure to compare the domination of two populations in a pair-wise manner, i.e., how well population x dominates population y, as well as how good population y *dominates* population x. This measure is defined as in Eq. 9.6:

$$C\left(N, N''\right) = \frac{\left|\left\{a'' \in N''; \exists a \in N : a a''\right\}\right|}{\left|N''\right|} \tag{9.6}$$

The value $C\ (N, N'') =$ means that all solutions in N are dominated by or equal to solutions in N''. $C\ (N, N'') = 0$ represents the situation when none of the solutions in N'' are covered by the set N. Note that both $C\ (N, N'')$ and $C\ (N'', N)$ have to be considered, since $C\ (N, N'')$ is not necessarily equal to $C\ (N'', N)$.

9.6 Experimental Results

To measure the performance of the algorithms, a benchmark simulation model (Braun et al. 2001) is considered. The simulation model is based on ETC matrix. As the heuristics are static, it is assumed that an accurate estimate of the expected execution time for each task in same time units on each processor is known prior to the execution and contained within an ETC matrix. Any ETC matrix will have $n*m$ entries, where n is the number of tasks and m is the number of processors. The ETC matrices are generated using the following method.

9.6.1 Data Set Description

Initially, an $n \times 1$ baseline column vector V is generated by repeatedly selecting m uniform random floating point values between 1 and φ_v, the upper bound on values in V. Each value $V\ (i)$ in V is multiplied by a uniform random number r which has an upper bound of φ_r. Each row in the ETC matrix is then given by $V\ (i) \times r$. One row requires m different values of r. This process is repeated for each row until all the rows are filled up. The vector V is not used in the actual matrix. Hence, the value in the ETC matrix is within the range $(1, \varphi_v, \varphi_r)$.

The characteristics of the ETC matrices are varied in an attempt to represent a range of possible heterogeneous environments. The variations are caused using three metrics: task heterogeneity, processor heterogeneity, and consistency. Task heterogeneity is defined as the amount of variance possible among the execution times of the tasks with two low and high possible values. Machine heterogeneity is the variation of the execution time of a particular task across all the processors which can be high and low. High task heterogeneity is represented by $\varphi_v = 3{,}000$ and low task heterogeneity uses $\varphi_v = 100$. A value of $\varphi_r = 1{,}000$ models high processor heterogeneity and $\varphi_r = 10$ models low processor heterogeneity. Three

Table 9.2 Parameter settings used in MOGA for scheduling independent tasks

Parameter	WSGA	NSGA-II	NSGA-II-CE	NSGA-II-PHC
Population size	200	200	200	200
Number of generations	2,000	1,000	1,000	500
Crossover probability P_c	0.8	0.8	0.8	0.8
Mutation probability P_m	0.02	0.02	0.02	0.02
Reduction rate r	–	–	0.65	–
Neighborhood size	–	–	–	5

different ETC consistencies, namely, consistent, inconsistent, and semiconsistent are used to model the features of a real heterogeneous system. An ETC matrix is considered consistent if a processor P_i executes task "*t*" faster than processor P_j, then P_i executes all the tasks faster than P_j. To model a consistent matrix, each row in the matrix is sorted independently with processor P_i always being the fastest and P_j being the slowest. Inconsistency means that a processor is faster for some tasks and slower for some others. Inconsistent matrices are left in the random state in which they are generated. A semiconsistent ETC matrix is characterized by an inconsistent matrix which has a consistent submatrix of a predefined size. Semiconsistent matrices are generated by extracting the even column elements of each row, sorting them, and replacing them while the odd column elements are left the same.

These different considerations combine to form 12 distinct types of possible ETC matrices, which simulate a range of different possible heterogeneous systems. The ETC matrices used comprise 512 tasks and 16 processors. The different problem instances are identified according to the following scheme:

$$x - yy - zz$$

where

x denotes the type of consistency (c – consistent, i – inconsistent, and s means semiconsistent).

yy indicates the heterogeneity of the tasks (hi – high and lo – low).

zz indicates the heterogeneity of the processors (hi – high and lo – low).

9.6.2 *Test Parameters of MOGA*

The proposed algorithms have been implemented using C language. All the experiments have been conducted on Pentium IV machines with 1GB of main memory loaded with Linux operating system. The algorithms are applied to the 12 benchmark instances simulated using the specifications given in Sect. 9.6.1. The parameters used in WSGA, NSGA-II, NSGA-II-CE, and NSGA-II-PHC are tabulated in Table 9.2.

The crossover operation is carried out with a higher probability and mutation is done at a reduced rate for better performance. The values of P_c and P_m tabulated are chosen by trial and error. A neighborhood size of 5 is chosen based on the number of objectives, generalizing to $2m+1$ where m is the number of objectives.

9.6.3 WSGA Test Results

The solutions obtained using WSGA after 2,000 iterations are evaluated against the weighted objective function for the two objectives make-span and flow-time. The values of make-span and mean flow-time measured in same time units obtained for the best schedule for all the 12 instances are given in Table 9.3. The Min–Min heuristic that outperforms the other heuristics (Braun et al. 2001) is implemented for comparing the performances. A single-objective GA that considers both the objectives separately is also implemented. The single-objective GA also uses an initial population of 200 solutions that are randomly generated, a single-point crossover, and a swap mutation with probability of 0.8 and 0.02, respectively. The algorithm is run for 2,000 generations for each of the objectives – the make-span and mean flow-time. The percentage of reduction in make-span and mean flow-time achieved by WSGA and GA over Min–Min is calculated as given by Eqs. (9.7) and (9.8):

$$\text{makespan reduction}\,(\%) = 1 - \frac{\text{makespan}_{\text{WSGA (GA)}}}{\text{makespan}_{\text{min-min}}} \tag{9.7}$$

$$\text{mean flowtime reduction}\,(\%) = 1 - \frac{\text{mean flowtime}_{\text{WSGA(GA)}}}{\text{mean flowtime}_{\text{min-min}}} \tag{9.8}$$

The results in Table 9.3 are analyzed for the minimization of two objectives: make-span and mean flow-time. It is observed that both GA and WSGA outperform Min–Min heuristic in terms of both the objectives. Comparing single-objective GA with WSGA it is found that WSGA achieves a make-span reduction of 26.9 % and mean flow-time reduction of 20.98 %. The scale of reduction achieved by GA for make-span and mean flow-time is 22.1 % and 18.3 %, respectively. This shows that WSGA produces optimal schedules comparatively.

9.6.4 Test Results of Pareto-Based MOGA

The other algorithms, namely, NSGA-II, NSGA-II–CE, and NSGA-II-PHC obtain a set of optimal solutions against the single solution obtained using WSGA. Figure 9.2 shows the plot of the Pareto-optimal solutions obtained using NSGA-II after running the algorithm for 200, 500, 700, and 1,000 generations, respectively.

Table 9.3 Performance of WSGA in scheduling independent tasks

Instance	Min–Min		GA		WSGA		Reduction by GA (%)		Reduction by WSGA (%)	
	Make-span MS (tu)	Mean flow-time MFT (tu)	Make-span MS (tu)	Mean flow-time MFT (tu)	Make-span MS (tu)	Mean flow-time MFT (tu)	MS	MFT	MS	MFT
c_lo_lo	9,349.31	119,659.34	9,604.34	108,042.41	**9,581.53**	**106,456.40**	18.6	9.9	18.9	11.03
c_lo_hi	844,392.3	9,472,765.5	542,992.10	6,721,964.29	**484,886.84**	**6,558,902.13**	35.6	29.0	42.5	30.76
c_hi_lo	288,690.4	3,596,294.9	233,113.29	3,200,261.62	**222,023.73**	**3,165,899.27**	19.2	11.0	23.09	11.96
c_hi_hi	25,012,959	281,268,501.1	15,650,312	201,879,847.6	**13,959,354**	**192,902,646.6**	37.4	28.2	44.19	31.41
s_lo_lo	10,984.30	138,113.37	8,397.67	115,988.98	**7,951.57**	**111,028.67**	23.5	16.0	27.6	19.61
s_lo_hi	1,141,802.6	11,969,049.52	683,059.31	8,889,134.39	**661,760.44**	**8,526,164.42**	40.1	25.7	42.04	28.76
s_hi_lo	317,131.51	4,057,957.29	267,827.96	3,401,507.475	**242,335.20**	**3,329,371.79**	15.5	16.1	23.58	17.95
s_hi_hi	29,818,693	358,289,060.7	20,793,029	268,727,377.7	**19,255,703**	**253,897,333.2**	30.2	24.9	35.42	29.13
i_lo_lo	10,722.18	139,211.47	8,917.90	122,493.90	**8,846.40**	**120,342.4152**	16.8	12.0	17.49	13.55
i_lo_hi	941,143.95	13,086,639.99	824,397.21	11,019,215.54	**804,858.26**	**10,256,117.40**	12.4	15.7	14.48	21.62
i_hi_lo	304,858.30	4,221,872.45	282,549.25	3,658,032.55	**264,065.21**	**3,570,443.152**	7.3	13.3	13.38	15.42
i_hi_hi	29,205,315	382,713,410.9	26,753,234	312,299,016.2	**23,342,636**	**303,613,189.0**	8.3	18.3	20.07	20.66
Average							22.1	18.3	**26.9**	**20.98**

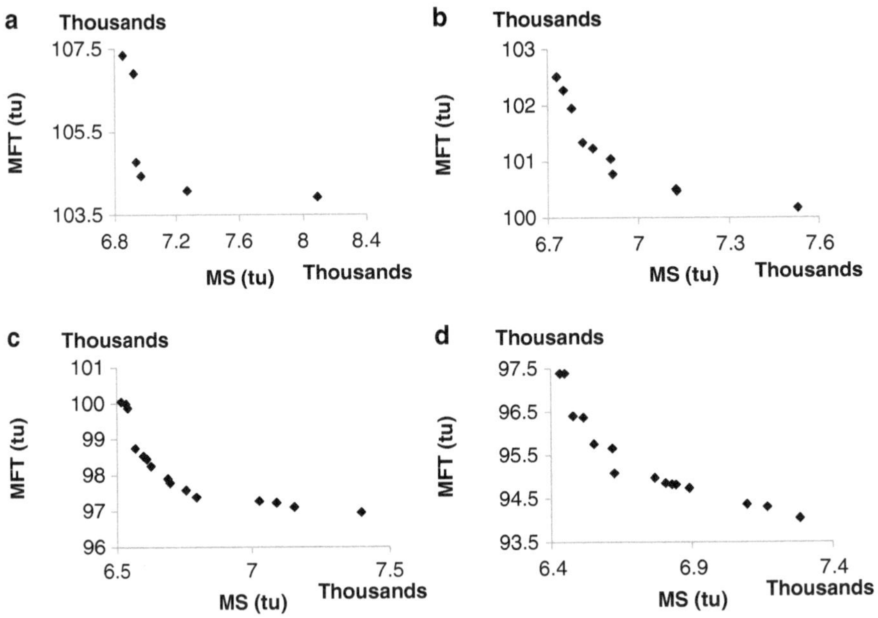

Fig. 9.2 Pareto-optimal solutions for the c_lo_lo instance using NSGA-II (**a**) After 200 iterations (**b**) After 500 iterations (**c**) After 700 iterations (**d**) After 1,000 iterations

The solutions are plotted for the consistent ETC matrices considering both low task and processor heterogeneity. The values of the objectives are scaled by 1,000 units for the Pareto-optimal solutions plotted.

From the solutions plotted in Fig. 9.2, it is seen that the number of Pareto-optimal solutions obtained increases as the number of iterations increases. It can also be seen that the quality of solutions improve as the number of iterations increases. This is indicated by the values of make-span and flow-time achieved. The algorithm is run on all the test instances to obtain the set of Pareto-optimal solutions.

For effective comparisons, NSGA-II-CE is also run for the same 1,000 iterations. The hybrid algorithm NSGA-II-PHC enhances the search by incorporating hill climbing at the end of each generation. As a result, solutions similar to NSGA-II and NSGA-II-CE are obtained by the hybrid method in lesser iterations. Hence, the hybrid method is terminated after running it for 500 iterations. The Pareto-optimal solutions obtained by the three methods are plotted in Fig. 9.3 for effective comparison.

From the results plotted in Fig. 9.3, it is found that Pareto-optimal approaches NSGA-II, NSGA-II-CE, and NSGA-II-PHC obtain a set of solutions against a single solution obtained using WSGA. The set of solutions obtained by the hybrid NSGA-II-PHC produces solutions that are nondominated by the set of solutions

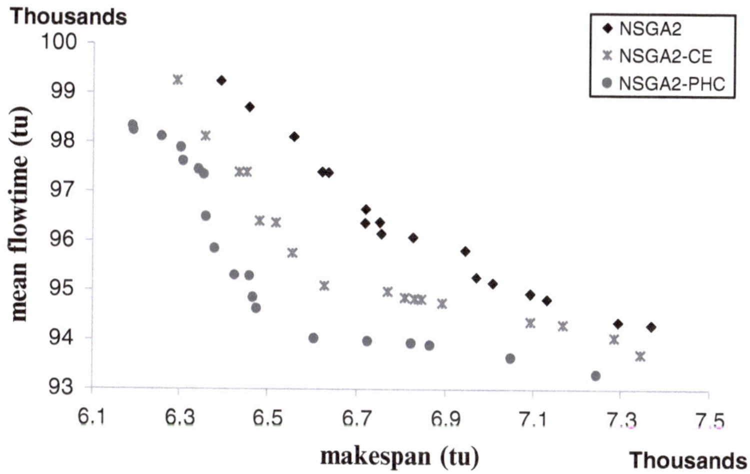

Fig. 9.3 Comparison of Pareto-optimal solutions for c_lo_lo instance obtained using Pareto-based MOGA

produced by the other two algorithms. To make an effective comparison of the weight-based and the Pareto-based MOGA methods, a fuzzy-based membership function is used to obtain a single best solution in the case of Pareto-based genetic algorithms. The values of make-span and mean flow-time of the best schedule obtained using the algorithms are compared in Table 9.4 for all the benchmark instances.

From the values of make-span and mean flow-time tabulated in Table 9.4, it is found that solutions obtained using NSGA-II-PHC minimizes both make-span and mean flow-time for all the instances. Also, the results obtained show that NSGA-II performs better schedules than WSGA. NSGA-II-CE performs over NSGA-II due to the enhancement in the selection process.

To determine the quality of the solutions, the algorithms are evaluated against a set of performance metrics described, namely, Ratio of Nondominated Individuals (RNI), Maximum Spread ($S_{\max}$), and Coverage of Two sets (C). The other performance metrics are not considered as they require knowledge of the true Pareto-optimal solutions that are not known for this problem. The results in Table 9.5 infer that the solutions obtained by NSGA-II-PHC are of high quality exhibiting enhanced performance in comparison with the other versions of NSGA-II for all types of instances with respect to the aforementioned quality measures.

It is observed that more number of Pareto-optimal solutions are obtained using NSGA-II-PHC from the value of 0.36 for *RNI* parameter. Achieving a distributed set of solutions minimizes the chances of local convergence and it is seen that NSGA-II-PHC covers a larger search space as indicated by a high value of $S_{\max}$. Further, the solutions produced by NSGA-II-PHC are dominated by the other solutions only to a very small extent, whereas they dominate the solutions produced by the other algorithms to a large extent, which is revealed from **C**.

Table 9.4 Comparison of weight-based and Pareto-based MOGA in scheduling independent tasks

	WSGA		NSGA-II		NSGA-II-CE		NSGA-II-PHC	
Instance	MS (tu)	MFT (tu)	MS (tu)	MFT (tu)	MS (tu)	MFT (tu)	MS (tu)	MFT (tu)
c_lo_lo	7,581.53	106,456.40	6,771.35	94,980.49	6,725.50	93,982.06	**6,392.70**	**93,510.04**
c_lo_hi	484,886.84	6,558,902.13	378,653.26	5,138,918.39	375,067.15	5,131,411.60	**366,134.28**	**5,093,236.14**
c_hi_lo	222,023.73	3,165,899.27	214,915.50	2,843,648.47	213,772.71	2,835,076.41	**211,687.44**	**2,723,601.84**
c_hi_hi	13,959,354.0	192,902,646.6	12,639,068.8	154,068,968.8	12,594,913.8	154,068,968.8	**12,498,541.1**	**145,386,657.5**
s_lo_lo	7,951.57	111,028.67	6,749.58	82,425.7044	6,256.56	81,384.66	**6,546.77**	**80,176.05**
s_lo_hi	661,760.44	8,526,164.42	460,817.50	5,282,076.98	457,470.24	5,264,674.90	**459,255.88**	**5,263,421.67**
s_hi_lo	242,335.20	3,329,371.79	186,560.16	2,371,762.68	182,063.70	2,347,563.76	**181,861.65**	**2,328,141.92**
s_hi_hi	19,255,703.6	253,897,333.2	12,855,976.4	159,959,678.5	12,849,014.8	158,286,572.9	**12,596,650.9**	**155,182,688.7**
i_lo_lo	8,846.40	120,342.41	6,451.19	79,886.54	6,159.96	77,898.01	**6,135.87**	**74,839.57**
i_lo_hi	804,858.26	10,256,117.4	525,548.58	5,838,798.12	485,256.58	5,794,749.75	**478,350.87**	**5,538,714.78**
i_hi_lo	264,065.21	3,570,443.15	182,241.14	2,355,441.18	180,760.14	2,348,690.00	**170,087.18**	**2,296,725.99**
i_hi_hi	23,342,636.9	303,613,189.0	13,852,895.8	176,479,328.3	13,566,648.2	163,574,373.3	**12,802,770.3**	**160,185,421.9**

Table 9.5 Performance comparison of Pareto-based MOGA in scheduling independent tasks

	RNI			S_{max}			Set Coverage (C)			
Instance	*NSGA-II*	*NSGA-II-CE*	*NSGA-II-PHC*	*NSGA-II*	*NSGA-II-CE*	*NSGA-II-PHC*	*NSGA-II over NSGA-II-CE*	*NSGA-II-CE over NSGA-II*	*NSGAII-CE over NSGA-II-PHC*	*NSGA-II-PHC over NSGA-II-CE*
c_lo_lo	0.150	0.180	**0.360**	3,457.05	4,900.86	**14,988.46**	0	0.80	0	**0.83**
c_lo_hi	0.100	0.170	**0.250**	263,522.02	531,825.3	**587,853.49**	0.23	0.80	0	**0.64**
c_hi_lo	0.140	0.160	**0.310**	127,956.76	132,127.9	**439,093.00**	0	0	0.07	**0.29**
c_hi_hi	0.180	0.120	**0.300**	10,104,059.6	19,152,222.9	**31,477,357.6**	0	0.94	0.11	**0.53**
s_lo_lo	0.060	0.080	**0.150**	2,364.71	4,459.10	**11,771.24**	0	1	0	**0.86**
s_lo_hi	0.040	0.100	**0.220**	104,004.42	207,827.1	**563,323.62**	0	1	0	**1**
s_hi_lo	0.060	0.090	**0.120**	74,651.66	95,653.68	**285,629.03**	0	1	0	**0.83**
s_hi_hi	0.050	0.060	**0.110**	6,765,694.0	10,044,344.4	**16,220,955.9**	0	1	0	**1**
i_lo_lo	0.070	0.090	**0.110**	2,172.27	4,693.19	**8,627.44**	0.22	0.42	0	**0.55**
i_lo_hi	0.030	0.050	**0.200**	172,170.80	497,344.9	**1,162,622.43**	0	0.40	0	**1**
i_hi_lo	0.050	0.080	**0.140**	80,164.33	89,624.62	**397,186.55**	0	1	0	**0.92**
i_hi_hi	0.070	0.070	**0.120**	7,173,656.9	12,118,347.7	**30,198,406.4**	0	1	0	**0.70**

Table 9.6 Parameter settings used in MOPSO for scheduling independent tasks

Parameter	WSPSO	NSPSO	ANSPSO	NSPSO-PHC
Population size	200	200	200	200
Number of generations	2,000	1,000	1,000	500
C_1	2	2	$C_1 i) = 2.5$	$C_1 i) = 2.5$
			$C_1(f) = 0.5$	$C_1(f) = 0.5$
C_2	2	2	$C_2 i) = 0.5$	$C_2 i) = 0.5$
			$C_2(f) = 2.5$	$C_2(f) = 2.5$
w_{max}	0.7	0.7	0.7	0.7
w_{min}	0.4	0.4	0.4	0.4
Neighborhood size	–	–	–	5

9.6.5 *Test Parameters of MOPSO to Schedule Independent Tasks*

This section discusses the results obtained using the proposed MOPSO algorithms for the independent task-scheduling problem. The parameters used in conducting the experiments using WSPSO, NSPSO, ANSPSO, and NSPSO-PHC are given in Table 9.6.

The initial population size and the number of generations are kept the same as in the case of GA for effective analysis. The values chosen for the parameters C_1, C_2, w_{max}, and w_{min} are from the literature (Li 2003; Yan Kang et al. 2013).

9.6.6 *Test Results of WSPSO*

The same instances used in analyzing the performance of MOGA methods are used to evaluate the MOPSO methods. As WSPSO uses an aggregating approach, the algorithm generates a single optimal solution for each instance considered. The values of make-span and mean flow-time obtained for the optimal solution at the end of 2,000 runs is given in Table 9.7.

From the results tabulated, it is observed that WSPSO produces schedules with higher values of make-span and mean flow-time for all the instances, though both WSPSO and WSGA work on the same initial population. WSGA achieves an average reduction of make-span and flow-time over WSPSO by 24.65 % and 20.13 %, respectively. The results also show that WSPSO produces solutions with large differences in the objectives for instances of high processor heterogeneity type than for other instance types.

Table 9.7 Performance of WSPSO in scheduling independent tasks

	WSGA		WSPSO		Reduction by WSGA (%)	
Instance	MS (tu)	MFT (tu)	MS (tu)	MFT (tu)	MS	MFT
c_lo_lo	**7,581.53**	**106,456.40**	9,061.29	117,340.49	16.33	9.27
c_lo_hi	**484,886.84**	**6,558,902.1**	800,174.51	9,448,456.4	39.40	30.58
c_hi_lo	**222,023.73**	**3,165,899.2**	280,472.76	3,508,367.3	20.83	9.71
c_hi_hi	**13,959,354.0**	**192,902,646.6**	23,717,256.9	280,113,008.3	41.14	31.13
s_lo_lo	**7,951.57**	**111,028.67**	10,861.21	135,950.51	26.78	18.33
s_lo_hi	**661,760.44**	**8,526,164.4**	1,045,787.69	11,908,505.4	36.72	28.40
s_hi_lo	**242,335.20**	**3,329,371.7**	315,714.63	4,042,059.73	23.24	17.63
s_hi_hi	**19,255,703.6**	**253,897,333.2**	28,917,091.9	356,144,150.1	33.41	28.70
i_lo_lo	**8,846.40**	**120,342.41**	10,164.29	137,990.510	12.96	12.78
i_lo_hi	**804,858.2**	**10,256,117.40**	941,143.95	13,010,902.20	14.48	21.17
i_hi_lo	**264,065.21**	**3,570,443.1**	302,129.04	4,181,565.18	12.59	14.61
i_hi_hi	**23,342,636.9**	**303,613,189.0**	28,445,995.1	376,070,915.9	17.94	19.26
Average					**24.65**	**20.13**

9.6.7 Test Results of Pareto-Based MOPSO Methods

The other algorithms – NSPSO, ANSPSO, and NSPSO-PHC – that are based on the Pareto-optimal approach produce a set of optimal solutions. Figure 9.4 shows the plot of the Pareto-optimal solutions obtained using NSPSO after 300, 500, 700, and 1,000 generations, respectively. The solutions are plotted for the consistent ETC matrix that models low task and low processor heterogeneity. The values of the objectives are scaled by 1,000 units for the Pareto-optimal solutions plotted.

The solutions plotted by running the algorithm for several iterations show that the number of Pareto-optimal solutions increases as the number of iterations increases. To improve the performance of NSPSO, the inertia weights and the learning factors are dynamically varied using the crowding distance parameter in ANSPSO. The algorithm is run for 1,000 generations. Further, the hybrid version of NSPSO, namely, the NSPSO-PHC is developed and the algorithm is stopped with 500 runs for effective comparisons. The solutions obtained by all the three algorithms are plotted in Fig. 9.5.

The comparison plot of the various Pareto-based MOPSO methods also show that the schedules produced by the hybrid method is better in terms of both make-span and flow-time for the considered test instance. Similar results are obtained for all the other test instances also. To compare Pareto-based methods with weight-based methods, a best compromise solution is selected from the set of solutions obtained by NSPSO, ANSPSO, and NSPSO-PHC. A fuzzy membership function is used in selecting the best compromise solution from the Pareto-optimal set. The make-span and the mean flow-time values thus obtained are tabulated in Table 9.8.

The results tabulated indicate that the schedule produced by NSPSO is better than WSPSO which is further enhanced by ANSPSO and the hybrid method

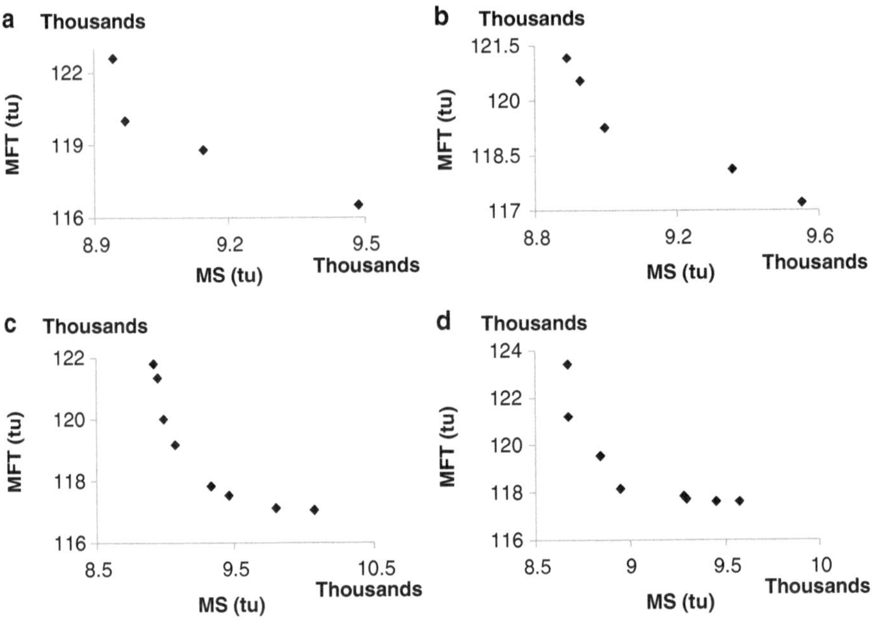

Fig. 9.4 Pareto-optimal solutions for c_lo_lo instance using NSPSO (**a**) After 300 generations (**b**) After 500 generations (**c**) After 700 generations (**d**) After 1,000 generations

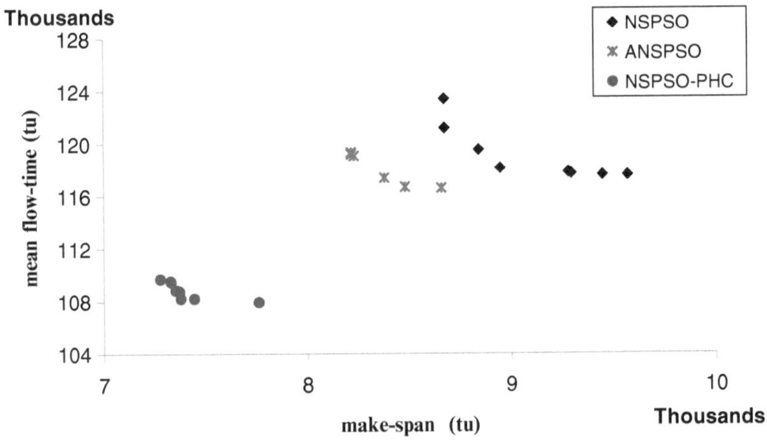

Fig. 9.5 Comparison of Pareto-optimal solutions for c_lo_lo instance obtained using Pareto-based MOPSO methods

Table 9.8 Comparison of weight-based and Pareto-based MOPSO methods in scheduling independent tasks

	WSPSO		NSPSO		ANSPSO		NSPSO-PHC	
Instance	MS (tu)	MFT (tu)	MS (tu)	MFT (tu)	MS (tu)	MFT (tu)	MS (tu)	MFT (tu)
c_lo_lo	9,061.29	117,340.49	8,664.80	117,821.78	8,481.28	116,687.00	**7,384.18**	**108,224.97**
c_lo_hi	800,174.51	9,448,456.40	779,897.71	9,403,574.69	650,590.82	8,791,838.84	**460,922.11**	**6,874,772.93**
c_hi_lo	280,472.76	3,508,367.36	277,143.87	3,488,426.44	263,132.55	3,382,033.24	**217,289.17**	**3,247,415.73**
c_hi_hi	23,717,256.9	280,113,008.3	23,572,329.5	278,368,048.6	18,525,774.9	248,456,121.4	**14,647,388.4**	**210,707,303.7**
s_lo_lo	10,861.21	135,950.51	10,842.05	135,609.86	10,476.29	133,918.63	**8,121.05**	**112,775.35**
s_lo_hi	1,045,787.6	11,908,505.49	1,015,471.54	11,646,285.23	960,156.32	11,565,127.32	**683,905.85**	**8,965,744.29**
s_hi_lo	315,714.63	4,042,059.73	308,564.65	4,017,087.20	302,462.87	3,993,035.73	**241,557.96**	**3,496,100.81**
s_hi_hi	28,917,091.9	356,144,150.1	28,165,967.0	348,987,256.0	27,706,329.7	340,515,495.8	**16,242,127.5**	**214,335,738.6**
i_lo_lo	10,164.29	137,990.51	10,141.6785	135,478.73	9,992.79	133,654.97	**8,569.72**	**116,200.05**
i_lo_hi	941,143.95	13,010,902.2	932,168.499	12,759,246.05	927,046.68	12,691,605.28	**581,280.54**	**7,605,271.72**
i_hi_lo	302,129.04	4,181,565.18	299,918.325	4,171,033.93	299,892.19	4,140,311.16	**207,706.37**	**2,911,375.02**
i_hi_hi	28,445,995.1	376,070,915.9	28,165,192.5	367,546,405.2	27,299,363.4	363,928,422.7	**18,716,797.0**	**237,609,881.9**

NSPSO-PHC outperforms the other two methods. A considerable amount of reduction in the considered objectives is obtained by NSPSO-PHC. This may be due to the mutation operation incorporated in the hill climbing process that alters the genes to a certain extent.

Further, the quality of solutions obtained by Pareto-based MOPSO methods are evaluated against the three metrics and the results obtained are tabulated in Table 9.9. The results of the Pareto-based MOPSO algorithms show that NSPSO-PHC produces larger number of solutions and covers a larger search area. The set of solutions are of high quality thereby dominating the entire set of solutions produced by the algorithms NSPSO and ANSPSO for all the instances.

9.6.8 Comparison of NSGA-II-PHC and NSPSO-PHC

Next, the two evolutionary methods GA and PSO applied for the task-scheduling problem are compared. Since the hybrid versions of both PSO and GA perform better, the Pareto-optimal solutions of both NSGA-II-PHC and NSPSO-PHC are plotted for the same instance in Fig. 9.6.

It is observed that NSGA-II-PHC produces solutions of very high quality, which results in minimum make-span and flow-time. Also, NSGA-II-PHC produces more number of solutions covering a larger search space as indicated by Table 9.10 for all the instances. However, the run time of the NSPSO-PHC algorithm is less compared to NSGA-II-PHC. This enables NSPSO-PHC achieve solutions quickly.

The comparison analysis for the quality of solutions produced by the hybrid MOGA and MOPSO methods indicate that NSGA-II-PHC produces more number of Pareto-optimal solutions, maximum number of solutions being achieved for the c_lo_lo instance as indicated by 0.36. For all the instances except i_ho_lo case, the set of solutions produced by NSGA2-PHC dominates the entire solution set of NSPSO-PHC. The CPU execution time of NSPSO-PHC algorithm is 59.23 s, which is comparatively less than 72.45 s, the time taken by NSGA-II-PHC.

9.7 Summary

In this chapter, multi-objective evolutionary algorithms based on GA and PSO are applied to schedule a set of independent tasks in an HDS. Both aggregating and Pareto-based approaches are used in solving the multi-objective optimization problem. The data set including instances of 512 tasks problems is simulated for testing the algorithms. The results for GA-based algorithms WSGA, NSGA-II, NSGA-II-CE, and NSGA-II-PHC are presented and analyzed. The schedules obtained by WSGA is better than both Min–Min heuristic and single-objective GA. WSGA shows a reduction in both make-span and mean flow-time values by 27 % and 21 % over Min–Min heuristic. The comparison of all GA-based algorithms shows

Table 9.9 Performance comparison of Pareto-based MOPSO methods in scheduling independent tasks

	RNI			S_{max}			Set Coverage (*C*)			
Instance	*NSPSO*	*ANSPSO*	*NSPSO-PHC*	*NSPSO*	*ANSPSO*	*NSPSO-PHC*	*NSPSO over ANSPSO*	*ANSPSO over NSPSO*	*ANSPSO over NSPSO-PHC*	*NSPSO-PHC over ANSPSO*
c_lo_lo	0.060	0.090	**0.120**	1,836.87	3,151.86	**4,457.99**	0	0.833	0	**1**
c_lo_hi	0.070	0.070	**0.110**	108,094.71	355,836.49	**392,659.9**	0	1	0	**1**
c_hi_lo	0.070	0.070	**0.110**	46,183.00	89,273.01	**124,934.7**	0	1	0	**1**
c_hi_hi	0.080	0.110	**0.120**	6,275,349.1	14,201,010.5	**28,244,178.1**	0	1	0	**1**
s_lo_lo	0.070	0.080	**0.150**	6,803.90	6,895.36	**8,534.29**	0	1	0	**1**
s_lo_hi	0.070	0.090	**0.090**	247,132.8	859,115.99	**1,056,967.9**	0	1	0	**1**
s_hi_lo	0.060	0.080	**0.100**	92,054.05	220,341.36	**238,319.99**	0	1	0	**1**
s_hi_hi	0.070	0.090	**0.110**	7,227,079.4	36,083,848.5	**41,320,603.2**	0	0.857	0	**1**
i_lo_lo	0.070	0.080	**0.100**	5,147.65	5,350.47	**6,396.46**	0	1	0	**1**
i_lo_hi	0.070	0.090	**0.110**	288,927.16	775,196.69	**1,099,626.41**	0	1	0	**1**
i_hi_lo	0.070	0.100	**0.140**	138,580.32	169,359.16	**245,987.75**	0	1	0	**1**
i_hi_hi	0.100	0.110	**0.130**	14,782,879.2	28,864,072.6	**29,508,440.4**	0	1	0	**1**

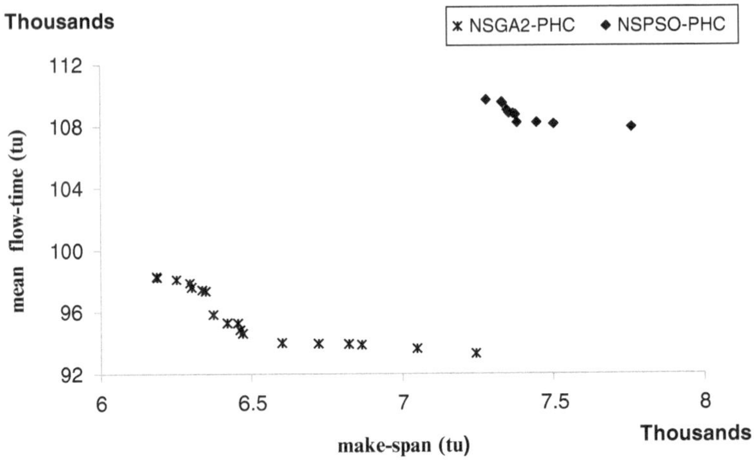

Fig. 9.6 Comparison of NSGA-II-PHC and NSPSO-PHC for c_lo_lo instance

Table 9.10 Performance comparison of hybrid methods NSGA-II-PHC and NSPSO-PHC

	RNI		S_{max}		Set Coverage (C)	
Instance	NSPSO-PHC	NSGA-II-PHC	NSPSO-PHC	NSGA-II-PHC	NSPSO-PHC over NSGA-II-PHC	NSGA-II-PHC over NSPSO-PHC
c_lo_lo	0.120	**0.360**	4,457.99	**14,988.46**	0	**1**
c_lo_hi	0.110	**0.250**	392,659.9	**587,853.49**	0	**1**
c_hi_lo	0.110	**0.310**	124,934.79	**439,093.00**	0	**1**
c_hi_hi	0.110	**0.300**	28,244,178.1	**31,477,357.6**	0	**1**
s_lo_lo	0.150	**0.150**	8,534.29	**11,771.24**	0	**1**
s_lo_hi	0.090	**0.220**	1,056,967.95	**563,323.62**	0	**1**
s_hi_lo	0.100	**0.120**	238,319.99	**285,629.03**	0	**1**
s_hi_hi	0.110	**0.110**	41,320,603.2	**16,220,955.9**	0	**1**
i_lo_lo	0.100	**0.110**	6,396.46	**8,627.44**	0	**1**
i_lo_hi	0.110	**0.200**	1,099,626.41	**1,162,622.43**	0	**1**
i_hi_lo	0.140	**0.140**	245,987.75	**397,186.55**	0	**0.428**
i_hi_hi	0.130	**0.120**	29,508,440.4	**30,198,406.4**	0	**1**

that NSGA-II-PHC produces good quality solutions and is effective when evaluated against the performance metrics for multi-objective problems. Similarly, the hybrid version of PSO, the NSPSO-PHC performs well among WSPSO, NSPSO, and ANSPSO algorithms. Comparison of MOGA and MOPSO methods indicate both weight based and Pareto-based approaches of MOGA perform better over MOPSO. WSGA shows a reduction in both make-span and mean flow-time values by 24 % and 20 % over WSPSO. Though NSPSO-PHC produces solutions quickly in comparison with NSGA-II-PHC, the solutions produced by NSGA-II-PHC are found to be more effective in terms of all the performance measures applied.

References

Abraham A, Liu H, Grosan C, Xhafa F (2008) Nature inspired meta-heuristics for grid scheduling: single and multi-objective optimization approaches, Studies in computational intelligence. Springer, Berlin/Heidelberg, pp 247–272

Ali S, Siegel HJ, Maheswaran M, Hensgen D, Ali S (2000) Representing task and machine heterogeneities for heterogeneous computing systems. Tamkang J Sci Eng 3(3):195–207

Braun TD, Siegel HJ, Beck N, Boloni LL, Maheswaran M, Reuther AI, Robertson JP, Theys MD, Yao B (2001) A comparison of eleven static heuristics for mapping a class of independent tasks onto heterogeneous distributed computing systems. J Parallel Distrib Comput 61(6):810–837

Carretero J, Xhafa F, Abraham A (2007) Genetic algorithm based schedulers for grid computing systems. Int J Innov Comput Inf Control 3(5):1053–1071

Coffman EG Jr (1976) Computer and job shop scheduling theory. Wiley, New York

Deb K (2001) Multi-objective optimization using evolutionary algorithms. Wiley, Chichester/New York

Eshaghian MM (1996) Heterogeneous computing. Artech House Publishers, Boston

Fleming PJ, Purshouse RC (2002) Evolutionary algorithms in control systems engineering: a survey. Control Eng Pract 10:1223–1241

Foster I, Kesselman C (2003) The grid: blueprint for a new computing infrastructure. Morgan Kaufmann, Amsterdam/Boston

Freund RF, Siegel HJ (1993) Introduction: heterogeneous processing. IEEE Comput Soc Press 26(6):13–17

Garey MR, Johnson DS (1979) Computers and intractability: a guide to the theory of NP-completeness. W. H. Freeman & Co, San Francisco

James HA (1999) Scheduling in metacomputing systems. PhD thesis, University of Adelaide, Australia

Krömer P, Abraham A, Snášel V, Platos J, Izakian H (2010) Differential evolution for scheduling independent tasks on heterogeneous distributed environments. Adv Intell Soft Comput 67:127–134

Lei Zhang, Yuehui Chen, Runyuan Sun, Shan Jing, Bo Yang (2008) A task scheduling algorithm based on PSO for grid computing. Int J Comput Intell Res 4(1):37–43

Li X (2003) A non-dominated sorting particle swarm optimizer for multi-objective optimization. In: Proceedings of Genetic and Evolutionary Computation Conference (GECCO'03) 2003, Chicago, IL, USA

Liu H, Abraham A, Hassanien A (2010) Scheduling jobs on computational grids using a fuzzy particle swarm optimization algorithm. Futur Gener Comput Syst 26(8):1336–1343

Page AJ, Keane TM, Naughton TJ (2010) Multi-heuristic dynamic task allocation using genetic algorithms in a heterogeneous distributed system. J Parallel Distrib Comput 70(7):758–766

Roy A, Das SK (2002) Optimizing qos-based multicast routing in wireless networks: a multi-objective genetic algorithms approach. In: Proceedings of the second IFIP-TC6 networking conference, Valencia, Spain

Subashini G (2013) Application of multi-objective evolutionary algorithms for task scheduling in heterogeneous distributed systems. PhD thesis, Anna University, India

Subashini G, Bhuvaneswari MC (2012) Comparison of multi-objective evolutionary approaches for task scheduling in distributed computing systems. Sadhana Acad Proc Eng Sci 37(6):675–694

Xhafa F, Carretero J, Abraham A (2007) Genetic algorithm based schedulers for grid computing systems. Int J Innov Comput Inf Control 3:1053–1071

Yan Kang, He Lu, Jing He (2013) A PSO-based genetic algorithm for scheduling of tasks in a heterogeneous distributed system. J Softw 8(6):1443–1450

Zitzler E, Deb K, Thiele L (2000) Comparison of multiobjective evolutionary algorithms: empirical results. Evol Comput 8(2):173–195

Author Index

M.C. Bhuvaneswari (ed.), *Application of Evolutionary Algorithms for Multi-objective Optimization in VLSI and Embedded Systems*, DOI 10.1007/978-81-322-1958-3,

Y

Z